Electrical Insulation Breakdown and Its Theory, Process, and Prevention:

Emerging Research and Opportunities

Boxue Du
Tianjin University, China

A volume in the Advances in Computer and Electrical Engineering (ACEE) Book Series

IGI Global
DISSEMINATOR OF KNOWLEDGE

Published in the United States of America by
IGI Global
Engineering Science Reference (an imprint of IGI Global)
701 E. Chocolate Avenue
Hershey PA, USA 17033
Tel: 717-533-8845
Fax: 717-533-8661
E-mail: cust@igi-global.com
Web site: http://www.igi-global.com

Library of Congress Cataloging-in-Publication Data

Names: Du, Boxue, 1961- author.
Title: Electrical insulation breakdown and its theory, process, and prevention : emerging research and opportunitites / by Boxue Du.
Description: Hershey, PA : Engineering Science Reference, [2019] | Includes bibliographical references.
Identifiers: LCCN 2019001914| ISBN 9781522588856 (h/c) | ISBN 9781522588863 (eISBN) | ISBN 9781522594574 (s/c)
Subjects: LCSH: Electric insulators and insulation--Defects. | Breakdown (Electricity)--Prevention.
Classification: LCC TK3401 .D822 2019 | DDC 621.319/37--dc23 LC record available at https://lccn.loc.gov/2019001914

This book is published in the IGI Global book series Advances in Computer and Electrical Engineering (ACEE) (ISSN: 2327-039X; eISSN: 2327-0403)

British Cataloguing in Publication Data
A Cataloguing in Publication record for this book is available from the British Library.

For electronic access to this publication, please contact: eresources@igi-global.com.

Advances in Computer and Electrical Engineering (ACEE) Book Series

ISSN:2327-039X
EISSN:2327-0403

Editor-in-Chief: Srikanta Patnaik, SOA University, India

MISSION

The fields of computer engineering and electrical engineering encompass a broad range of interdisciplinary topics allowing for expansive research developments across multiple fields. Research in these areas continues to develop and become increasingly important as computer and electrical systems have become an integral part of everyday life.

The **Advances in Computer and Electrical Engineering (ACEE) Book Series** aims to publish research on diverse topics pertaining to computer engineering and electrical engineering. **ACEE** encourages scholarly discourse on the latest applications, tools, and methodologies being implemented in the field for the design and development of computer and electrical systems.

COVERAGE

- Qualitative Methods
- Computer Hardware
- Circuit Analysis
- Chip Design
- Algorithms
- Computer Architecture
- Digital Electronics
- Computer Science
- Analog Electronics
- Applied Electromagnetics

The Advances in Computer and Electrical Engineering (ACEE) Book Series (ISSN 2327-039X) is published by IGI Global, 701 E. Chocolate Avenue, Hershey, PA 17033-1240, USA, www.igi-global.com. This series is composed of titles available for purchase individually; each title is edited to be contextually exclusive from any other title within the series. For pricing and ordering information please visit http://www.igi-global.com/book-series/advances-computer-electrical-engineering/73675. Postmaster: Send all address changes to above address.

Titles in this Series

For a list of additional titles in this series, please visit:
https://www.igi-global.com/book-series/advances-computer-electrical-engineering/73675

Handbook of Research on Emerging Applications of Fuzzy Algebraic Structures
Chiranjibe Jana (Vidyasagar University, India) Tapan Senapati (Padima Janakalyan Banipith (H.S.), India) and Madhumangal Pal (Vidyasagar University, India)
Engineering Science Reference • ©2020 • 300pp • H/C (ISBN: 9781799801900) • US $285.00

Challenges and Applications for Implementing Machine Learning in Computer Vision
Ramgopal Kashyap (Amity University, Raipur, India) and A.V. Senthil Kumar (Hindusthan College of Arts and Science, India)
Engineering Science Reference • ©2020 • 293pp • H/C (ISBN: 9781799801825) • US $195.00

Handbook of Research on Recent Developments in Electrical and Mechanical Engineering
Jamal Zbitou (University of Hassan 1st, Morocco) Catalin Iulian Pruncu (Imperial College London, UK) and Ahmed Errkik (University of Hassan 1st, Morocco)
Engineering Science Reference • ©2020 • 553pp • H/C (ISBN: 9781799801177) • US $255.00

Architecture and Security Issues in Fog Computing Applications
Sam Goundar (The University of the South Pacific, Fiji) S. Bharath Bhushan (Sree Vidyanikethan Engineering College, India) and Praveen Kumar Rayani (National Institute of Technology, Durgapur, India)
Engineering Science Reference • ©2020 • 205pp • H/C (ISBN: 9781799801948) • US $215.00

Handbook of Research on Advanced Applications of Graph Theory in Modern Society
Madhumangal Pal (Vidyasagar University, India) Sovan Samanta (Tamralipta Mahavidyalaya, India) and Anita Pal (National Institute of Technology Durgapur, India)
Engineering Science Reference • ©2020 • 591pp • H/C (ISBN: 9781522593805) • US $245.00

Novel Practices and Trends in Grid and Cloud Computing
Pethuru Raj (Reliance Jio Infocomm Ltd. (RJIL), India) and S. Koteeswaran (Vel Tech, India)
Engineering Science Reference • ©2019 • 374pp • H/C (ISBN: 9781522590231) • US $255.00

For an entire list of titles in this series, please visit:
https://www.igi-global.com/book-series/advances-computer-electrical-engineering/73675

Table of Contents

Preface

CONTEXT AND PURPOSE

With the continuous development of power system and the continuous improvement of power supply reliability, the power system has undergone unprecedented changes. In order to realize long-distance and large capacity transmission of energy, the original high-voltage AC transmission is transmitted by UHVDC; in order to realize the convenience of power supply system, the transformation from traditional overhead line to cable and Gil transmission is realized. With the improvement of the level of electrical engineering construction, the number of scientific and technological innovation is increasing year by year. The role of electrical engineering construction in modern industry, especially in the operation and management of power grid construction is becoming more and more important.

Modern power system is becoming more and more complex, which requires further improvement of power system reliability. The insulation problem of power system greatly limits the reliability improvement, and many researchers have made a lot of research on it. At this point, the combination of new technology, new material technology and production process is more rapid. The optimization design of insulation structure, basic theoretical research, simulation combined with experimental research have been in-depth development. It is particularly noteworthy that the in-depth discussion of these problems is conducive to the continuous improvement of power system reliability, and the theoretical and experimental research of insulation materials is conducive to fault analysis and insulation optimization. In the actual operation, the insulation of electrical equipment bears the comprehensive effect of electrical, thermal and mechanical stress. When the insulation failure occurs, it is not only caused by a certain problem, but also the result of multi physical field coupling, which makes the problem analysis extremely complex.

At the same time, in the design, installation and operation, any part may have insulation defects. The emerging detection technology can also fully avoid these problems. Therefore, the research on electrical equipment insulation, electrical insulation breakdown and its mechanism, process and prevention are worthy of study.

In essence, the breakdown, discharge, surface charge, dendrite structure, trap characteristics, charge transfer, surface charge and so on are important characteristics of insulating materials. The research on these characteristics has always been an important topic, especially in material modification. This provides a promising way to discover the new properties of polymer dielectrics. However, polymer dielectric is an interdisciplinary field involving chemistry, physics and mechanics. For academic researchers and industrialists, many principles and theories have not been confirmed, even unknown. Therefore, we write this book to provide some valuable research, hoping to inspire readers to new understanding and new research methods of polymer properties.

CHALLENGE

It goes without saying that the insulation problem of power system is always related to the reliability operation of power system. In order to improve the reliability of power system, many works have been done in the design of insulation structure and the development and modification of insulation materials. In recent years, with the gradual application of multi physical field finite element simulation, many problems have been fully solved. However, the problem of insulation itself is ultimately attributed to the basic properties of materials, such as dielectric constant, conductivity, dielectric loss, breakdown field strength, flashover voltage and so on. In order to improve the performance of these insulating materials, the solutions are gradually found through the research of nano adding dielectric, composite dielectric, spatial dielectric characteristic structure, electrochemical damage mechanism, micro dielectric phenomenon, space charge suppression and so on. But with the development of time, we still face many problems.

1. From the macroscopic point of view, whether the modified dielectric material affects the original breakdown, flashover, discharge and other properties;
2. From the microscopic point of view, after optimization, does the electrical tree growth characteristics, space charge characteristics, trap

characteristics, charge transport characteristics, surface charge and so on of dielectric insulating materials change?

3. The problem of uniform electric field in electrical equipment.
4. For the relationship between the two, the change of the dielectric macro electrical parameters is affected by those micro factors. In turn, whether the change of the micro factors significantly affects the dielectric macro electrical performance.
5. Whether new dielectric materials can be developed according to the macroscopic and microscopic characteristics of insulation.
6. Generally, the laboratory tests are under ideal conditions. Can we make the conditions consistent with the actual conditions?
7. Can the new dielectrics be widely used in engineering? What are the problems and how to improve them?

SEARCHING FOR A SOLUTION

In order to improve the reliability of power system operation, the discipline of high voltage and insulation technology takes on a very important task. From the theoretical level, the impact of power system failure is usually more serious, we should try to avoid the generation of power system failure. From the perspective of ontology, the purpose of reliability improvement can be realized by fully studying and improving the insulating materials. Therefore, there are many ways to improve the above problems.

1. Understand the breakdown and discharge phenomenon of advanced insulating materials through test
2. Analyze the insulation problems by measuring the micro electrical properties of the insulation materials;
3. Material modification is realized by connecting the microscopic and macroscopic properties of insulating materials;
4. Through the new design scheme, the problem of uniform electric field is realized.
5. Study on the characteristics of electrical tree to solve the problem of dielectric failure mechanism
6. Research on charge transport and dielectric properties to realize the research on conductivity and loss of insulating materials.

OVERVIEW OF CHAPTERS

This book is a collection of our previous researches. People with some basic background knowledge may find it with abundant content covering many hot topics in recent years. The book is organized into twelve chapters. A brief description of each of the chapters follows:

Chapter 1 identifies the existing collection of breakdown and discharge research on advanced materials in the new insulating material. In this chapter, epoxy resin is selected as the insulating material, and the nano materials with different content are modified to study the DC conductivity, space charge behavior, trap energy level distribution and breakdown field strength of the new material. In particular, we also discussed the mechanism of the impact.

Chapter 2 establishes the request for a GIL insulator that may occur the problem of flashover. The authors of this chapter contend that by experiment and simulation. Through the material spraying treatment of insulator surface, the change of dielectric parameters is realized, the electric field on insulator surface is more uniform, and the optimization of insulator volume is realized.

Chapter 3 takes HTV Silicone Rubber as research target, which is widely used in cable. Through the study of different pulse voltage, different temperature, the mechanical properties of silicone rubber changes, resulting in changes in electrical properties. To achieve the destructive analysis of the actual operating conditions.

Chapter 4 reviews the oil paper insulation defect of the converter transformer appears. We discuss the electric field and breakdown problem in the compound electric field. The results show that with the increase of temperature, the trap energy level and the number of shallow traps increase under DC voltage and compound voltage. This makes breakdown and flashover more likely.

Chapter 5 presents new problems of new environmental protection insulating material PE. Although PE is environmentally friendly, its electrical performance is not enough to meet the leading industry standards. Therefore, in this chapter, PE is modified reasonably to improve the electrical performance of PE. Through the measurement of DC conductivity, breakdown strength, trap energy level distribution, space charge distribution and charge mobility, the interpretation of the results is realized.

Chapter 6 addresses the space charge problem of traditional insulating material polyethylene. Through the improvement of PE and stabilizer modified by graphene oxide (GO) nanoparticles, the space charge can be suppressed and the electric field in the cable can be homogenized. We use differential scanning calorimetry (DSC) test, tensile test, breakdown test and conductivity test, and then use PEA method to measure the space charge behavior, and realize the explanation and mechanism analysis of space charge change.

Chapter 7 analyses the influence of pulse amplitude and frequency on PP tree at different temperatures. In addition, the effects of DC voltage, pulse voltage and pulse frequency on the tree shape characteristics of PP under DC voltage and pulse combination voltage were studied.

Chapter 8 discusses cable accessories are the most vulnerable locations in the cable system. Especially for DC cables, serious problem gradients will cause potential migration and field distortion. Therefore, we study the nonlinear conductivity of SIR/ SiC composite. At the same time, the influence of temperature on electric field is considered under the condition of DC voltage and DC superimposed pulse voltage.

CONCLUSION

While writing this book, we not only try to learn mistakes from the past, but also draw the scientific development and research fronts to it. Pay attention to spark inspiration and innovation of reader. On the whole, electrical insulation breakdown and its theory, process, and prevention are under consideration. On detail, from the study of basic properties of insulating materials to the research of electrical behavior and material modification. For example, the growth of electrical tree, charge behavior and discharge phenomena in different insulation system are investigated, and the modification of insulating materials is deeply analyzed. Of particularly note is that this book gives readers the guidance of research method, moreover, it provides suggestions for further research. Due to my limited knowledge, there may be a lot of inappropriateness in this book and I sincerely hope that readers will criticize and correct it.

Chapter 1
Collection of Breakdown and Discharge Research on Advanced Materials

ABSTRACT

Epoxy resins are widely used to build insulators in GIL. Epoxy/AlN nanocomposite can be produced by adding AlN nanoparticals to the epoxy resin. By studying the surface discharge behavior of the nanocomposites under different operating temperature, it is helpful to improve the creeping voltage of epoxy resin. Polypropylene is a kind of material which is usually chosen to build film capacitors. The effects of voltage form on surface charge and discharge behavior were studied. Furthermore, a modification method of a polypropylene film which can suppress surface charge accumulation is proposed. Polypropylene also has great application potential in HVDC cable insulation, provided that its toughness is to be overcome. Different mass fractions of ULDPE and graphene were added to polypropylene to improve mechanical and insulating properties, respectively. Studies on the DC conductivity, space charge behavior, trap level distribution, and breakdown strength of the new material were carried out.

DOI: 10.4018/978-1-5225-8885-6.ch001

INTRODUCTION

In the fields of urban power supply and electrical power delivery, high-voltage direct current (HVDC) transmission is widely used (Du, 2017). The advantages of gas insulated transmission line (GIL) include small footprint, high operational reliability and environmental friendliness compared to overhead lines (Zhang, 2017). An epoxy resin is used to make an insulator therein in a gas insulated pipeline transmission system. At DC voltage, the voltage distribution across the insulator depends on the distribution of its conductivity. When overload operation or partial discharge occurs, the operating temperature of the insulator rises, which causes a decrease in insulation performance of the insulator (Du, 2013). During the operation of gas insulated pipelines, creeping discharge is a serious problem. Compared with pure gas breakdown or breakdown of insulating material body, the voltage of creeping discharge is lower, and one of the important reasons for flashover occurs is surface charge concentration (Shao, 2017). The concentration and dissipation behavior of surface charge is affected and decided by the trap distribution characteristics in the materials. At the same time, related research shows that the flashover characteristics of the surface are also closely related to the charge trap (Li, 2010). Therefore, studying the dielectric behavior of epoxy resin under temperature rise is of great significance for improving the operational reliability of gas insulated transmission pipelines. With the development of nanotechnology, many scholars have studied the charge and discharge behavior of epoxy nanocomposites. Studies have found that the insulation properties of materials at higher temperatures are worse than at room temperature (Du, 2016). As the temperature increases, the charge dissipates faster and the traps are deeper and less (Du, 2017). Properly adding nanoparticles helps introduce new traps to capture free charges to achieve the effect of increasing the flashover voltage. However, what will happen if adding nanoparticles and under high temperature environments to the surface charge and flashover behavior of materials remains unknown. Aluminum nitride nanoparticles, a kind of material having high thermal conductivity and low dielectric constant at the same time, were selected as representative particles to be added to the epoxy resin. The effect of adding nanoparticles and different ambient temperature on the surface charge behavior and trap distribution of epoxy resin was studied by surface potential decay method, and then the flashover characteristics of epoxy nanocomposites were studied under the same environment. Through this research, it is helpful to find a suitable formula to improve the creeping voltage of the epoxy resin in different working temperature environments.

Film capacitors widely used in flexible DC transmission systems have excellent self-healing properties, corrosion resistance and low inductance. A film capacitor for a voltage source converter (VSC) is subjected to a DC voltage under normal operating conditions, but a transient overvoltage caused by a power system fault will act on the capacitor (Florentzou, 2009). In this case, the creeping flash in the film capacitor will be caused by the combined voltage of DC and pulse. The electrodes of the film capacitor are divided into small portions which are connected by a fusion bridge. If a local failure occurs, the bridge will melt and isolate the faulty part (Grafton, 2001). Flashover in film capacitors is one of the main causes of failure (Hole, 2001). A composite voltage consisting of a direct current and a pulse voltage causes an increased likelihood of flashover between the electrodes. Compared with other polymer films, polypropylene films are widely used to make film capacitors due to their low breakdown resistance, low dielectric loss factor, and ease of processing (Tsunoda, 2000). The electrical properties of the polypropylene film directly determine the reliability of the film capacitor. The trap in the polymer dielectric can capture charge under voltage, and the application of voltage over a long period of time will cause charge accumulation on the surface of the material (Wang, 2004). The presence of surface charge can cause localized non-uniformities in the electric field distribution, accelerating the electrical aging of the polymer and causing flashover (Kumara, 2012). The surface charge behavior becomes more complicated under the action of a composite voltage consisting of DC and pulse voltage. The accumulation of surface charge affects the flashover voltage along the surface. Therefore, it is important to study the development of flashover of polymer insulation under the action of surface charge on the safety and stability of film capacitors, the concentration distribution of surface charge suppression and the improvement of electric strength. Many scholars have studied the surface charge inhibition mechanism of dielectric polymers. Studies have shown that direct fluorination can optimize the surface state of the polymer and is an effective surface modification method that can change the surface charge accumulation characteristics. The formation of c-f bonds in the surface molecules of the polymer directly affects the dynamic properties of the charge on the surface and inside of the material. In order to suppress the accumulation of surface charge, it is necessary to grasp the charge transport law of the polymer. Carrier mobility is one of the important parameters of charge behavior and is closely related to charge transport. The carrier mobility of the polymer can be obtained by observing the electrical parameters exhibited in the dynamic process of the surface charge, and the experiment is simple and convenient to measure.

Polypropylene (PP), which has attracted much attention in the field of high voltage DC cable insulation research, has excellent dielectric properties and no cross-linking and high temperature resistance (Huang, 2017). However, in further practical applications, the main disadvantage of the inherent brittleness of PP must first be overcome. In response to this shortcoming, related research proposes blending PP with a thermoplastic elastomer, which is an effective means to effectively improve the mechanical properties of polypropylene (Green, 2015). In recent years, it has been found that the addition of ultra low density polyethylene (ULDPE) to rigid substrates is also a means of improving the mechanical properties of materials (Khare, 2000), which can be used to improve the mechanical properties of PP cable insulation. The space charge accumulation behavior under the strong DC electric field is still an important issue to be considered in the research of high-voltage DC cable polymer insulation. The accumulation of charge causes electric field distortion, accelerates insulation aging, and even leads to insulation breakdown failure (Du, 2017; Montanari, 2011). A large amount of research work shows that the doping of nanoparticles can inhibit the accumulation of space charge in the polymer insulation layer, because the interface region between the nano-filler and the polymer dielectric can prevent the movement of free charges and limit the migration of charge carriers., thereby suppressing the injection and transport of space charges (Tanaka, 2005; Li, 2015). Generally speaking, ordinary nano-fillers are theoretically difficult to obtain a large specific surface area. In order to achieve the desired effect, it is necessary to dope a large amount of nano-filler, which causes the particles to agglomerate and even cause concentration defects in the matrix. In recent years, graphene nanoplates and graphene-based composite materials having a special structure of a single atom thick layer and having a large specific surface area have been widely concerned. A small amount of graphene nanoplates can form a large number of interaction zones in the polymer, which is beneficial to further develop the potential of nano-dielectrics. Studies have found that the introduction of graphene brings many deep traps to polyethylene, inhibiting the accumulation of space charge. Studying the insulation properties of polypropylene composites with a small amount of graphene nanoplates will facilitate the development of high voltage DC cable insulation.

TEMPERATURE DEPENDENT SURFACE CHARGE AND DISCHARGE BEHAVIOR OF EPOXY/ALN NANOCOMPOSITES

Properties of Test Specimens

The samples used in the experiment were prepared by filling different amounts of nano-aluminum nitride into epoxy resin. The spatial distribution of nanoparticles in the epoxy resin composite can be observed by scanning electron microscopy (SEM). A cross-sectional scanning electron microscope image of an additive-free epoxy resin and an epoxy resin to which aluminum nitride nanoparticles are added is shown in Figure 1. A sample having a filler content of 1% by weight has good particle dispersibility. Based on the multi-core model of the polymer nanocomposite dielectric, it is known that

Figure 1. SEM images of the expoxy/AIN nanocomposites with filler contents of (a) 0 wt%, (b) 1 wt%, (c) 3 wt% and (d) 5 wt%

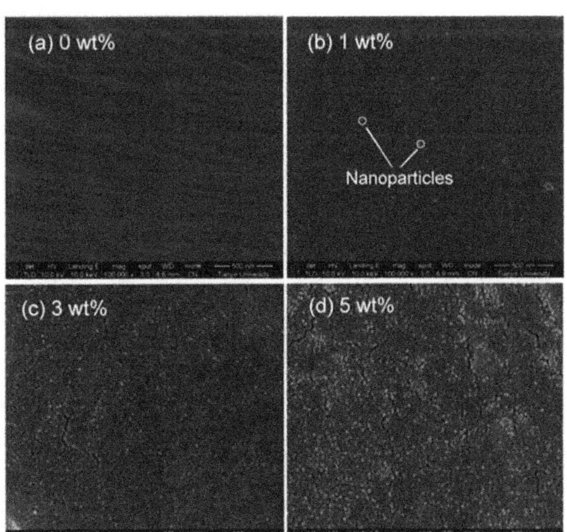

more filler content leads to a smaller pitch between the nanoparticles. The morphology of the interaction zone affects the chain conformation, chain mobility and free volume of the composite. To simplify the narrative, this effect is hereinafter referred to as agglomeration.

The relative dielectric constants of samples with different nanoparticle contents at different ambient temperatures at 50 Hz are shown in Figure 2. The relative dielectric constant of the nanocomposite is higher than that of the epoxy without the addition of nanoparticles, because the nano-AlN as an additive has a higher dielectric constant (Du, 2018) than the epoxy resin, so the relative dielectric constant The amount of addition increases and increases. In addition, the relative dielectric constant of the composite increases with increasing ambient temperature, and the higher the nanoparticle mass fraction, the greater the relative dielectric constant of the sample increases with temperature, indicating that the increase in ambient temperature is increased. The polarization ability of the sample, and with the increase of the nano-AlN content, the environmental temperature has an enhanced effect on the polarization of the composite.

Under the action of a high voltage, a large amount of electric charge flows through the sample to generate an electric current. The volumetric conductivity of the samples with different filler contents at ambient temperatures of 20°C and 80°C is shown in Figure 3. As the filler content increases, the surface conductivity decreases first and then increases. Among all the samples, the volume conductivity of the sample added with 1 wt% was the lowest. The volumetric conductivity of the epoxy resin decreases as the nanoparticles are added. The filler in the insulating material hinders the migration of carriers

Figure 2. Relationship between the relative permittivity at 50 Hz and the temperature

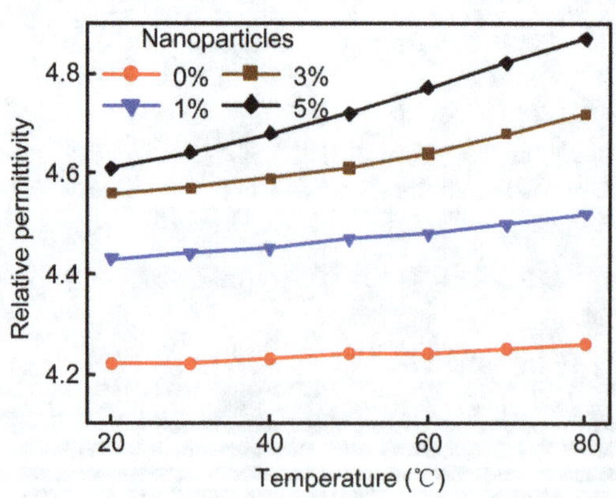

Figure 3. Relationship between the volume conductivity and the filler content at 20 and 80 °C

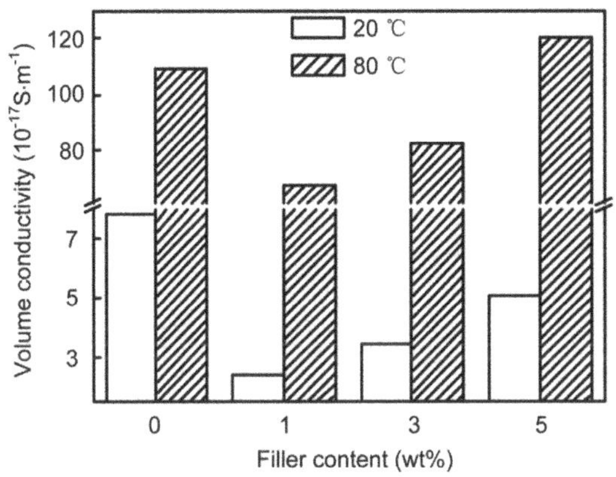

in the polymer. However, as the distance between the nanoparticles increases and gets closer as the amount of addition, the agglomeration phenomenon may form a channel for free charge migration. Therefore, for the epoxy resin to which the nanoparticles are added, the volumetric conductivity of the sample increases as the filler content increases. For experimental data at 80°C ambient temperature, the volumetric conductivity of the original epoxy sample is an order of magnitude higher than at 20°C. Higher ambient temperatures give the carrier higher energy for accelerated migration. At 80°C, the volume conductivity of the sample to which 1 wt% of the nanoparticles was added was also the lowest. The sample with the addition of 5 wt% aluminum nitride nanoparticles exhibited higher volume conductivity than the epoxy sample without the addition of nanoparticles, indicatingthat the increase in ambient temperature also accelerates the migration of electrons in the agglomeration region of the nanoparticles. In this case, a higher amount of addition contributes to the migration of carriers.

Figure 4 shows the surface conductivity of samples with different loadings at different temperatures. For the case where the ambient temperature is 20°C, the surface conductivity of the sample changes with the addition amount of the nanoparticles to be the same as the volume conductivity. The presence of nanoparticles can increase the travel required for carriers to pass through the surface of the insulating material. However, a higher filler content causes agglomeration, which leads to a decrease in the barrier effect of the nanoparticles on carrier migration and an increase in surface conductivity. At 80°C, the surface conductivity of the sample showed a tendency to differ from

Figure 4. Relationship between the surface conductivity and the filler content at 20 and 80 °C

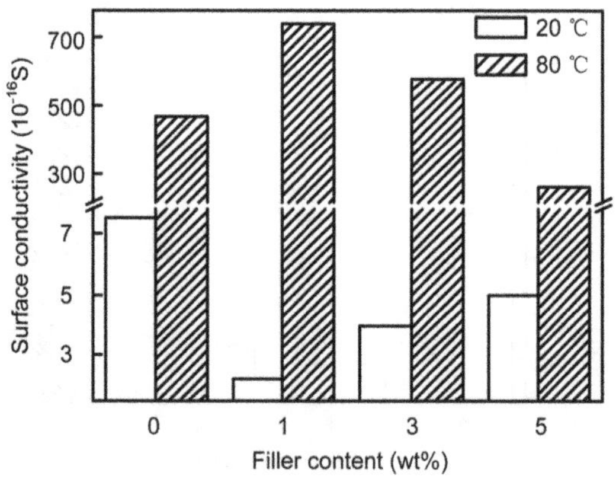

that at room temperature. This phenomenon is caused by the distribution of different shallow traps and deep traps in the material caused by the addition of different nanoparticles at 80°C.

Effect of Ambient Temperature

Since the charge dissipates faster at 80°C and the surface charge density quickly reaches zero, the surface charge dissipation process from 0 to 120 seconds is selected as shown in Figure 5. The decay of the surface potential versus time reflects the attenuation of the central charge density at the surface of the sample. The initial surface charge density of the sample charged at 20°C was about twice as high as that at 80°C. After 120 seconds, the surface charge at 80°C disappeared completely, and the surface potential at 20°C did not change much during this time. This is because when the temperature rises, the charge movement speed increases, and the charge trapped by the trap decreases, causing the initial surface charge density to decrease as the ambient temperature increases. At the same time, the charge dissipation rate also increases as the ambient temperature increases. This is due to the intense thermal motion that causes the charge to collapse faster.

Figure 6 shows the trap distribution of composite samples at different ambient temperatures. The resulting data is calculated experimentally based on the surface potential decay process. At 20°C, the distribution curve of the trap level and density of the original epoxy sample peaked around 0.8

Figure 5. Relationship between the surface charge density and the decay time of the specimens with filler contents of (a) 0 wt% and (b) 5 wt%

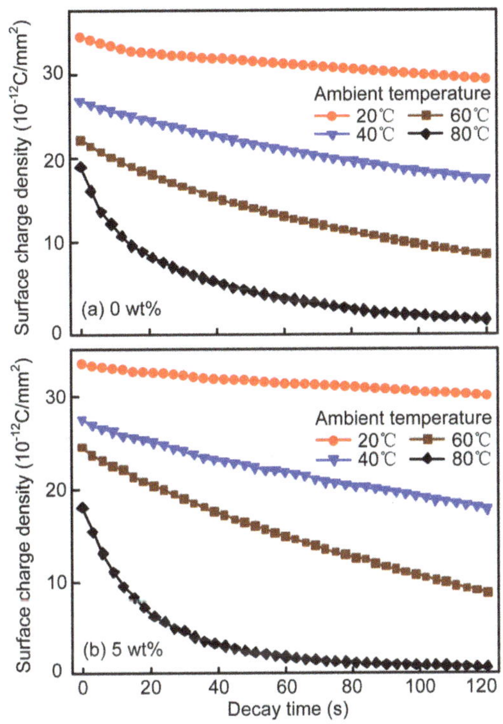

eV. The trap level corresponding to the highest point of the curve increases as the ambient temperature increases. For the sample with a filler content of 5%, the peak at 20°C appears at around 0.83 eV, while for the ambient temperature of 80°C there are two peaks, and the peak of the shallow trap appears at around 0.85 eV. The peak appears around 0.97ev. The results show that as the ambient temperature increases, the trap level deepens and the trap density decreases. The original epoxy and epoxy nanocomposites showed the same trapdistribution variation at both ambient temperatures. High ambient temperatures have two effects: first, the charges trapped by a higher-energy trap gets higher energy and collapses; second, the charge trapped by the trap is reduced after the charging process. The above reasons lead to higher trap levels and less trap density in high temperature environments.

Figure 6. Tap distribution of the specimens with filler contents of (a) 0 wt% and (b) 5 wt% at different temperature

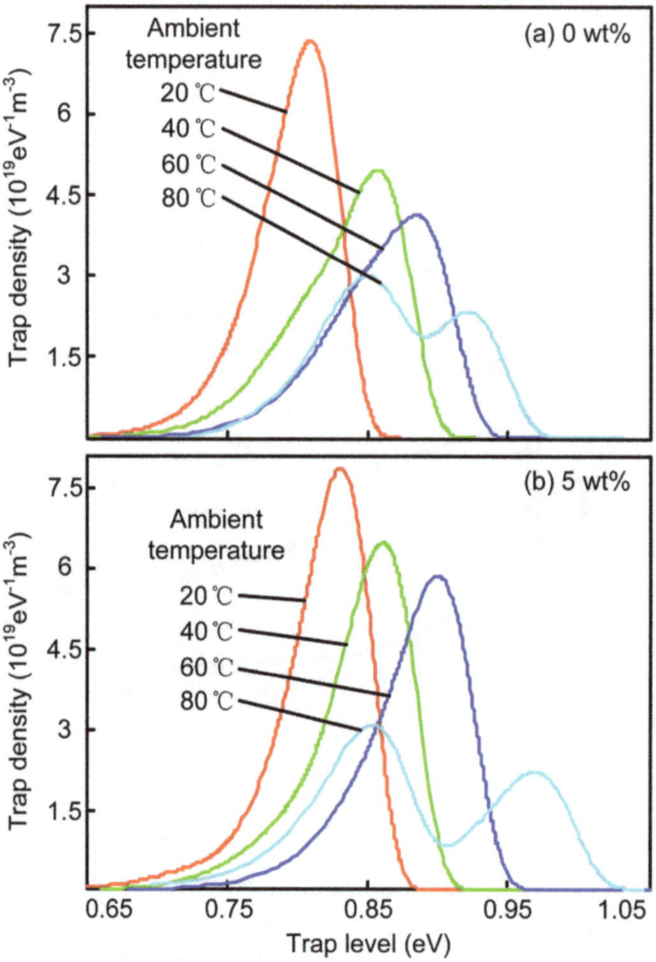

The Weibull plot of the original epoxy sample and the sample containing 5 wt% nano-AlN particles under the application of DC voltage to the electric field intensity corresponding to the surface flashover is shown in Figure 7. Figure 8 shows the electric field strength of the sample under flashover in different temperature environments under DC voltage. The data in the figure is the creeping field strength corresponding to the 63.2% discharge probability under the Weibull distribution. The epoxy resin sample to which

Figure 7. Weibull plots of the surface discharge strength of the specimens with filler contents of (a) 0 wt% and (b) 5 wt%

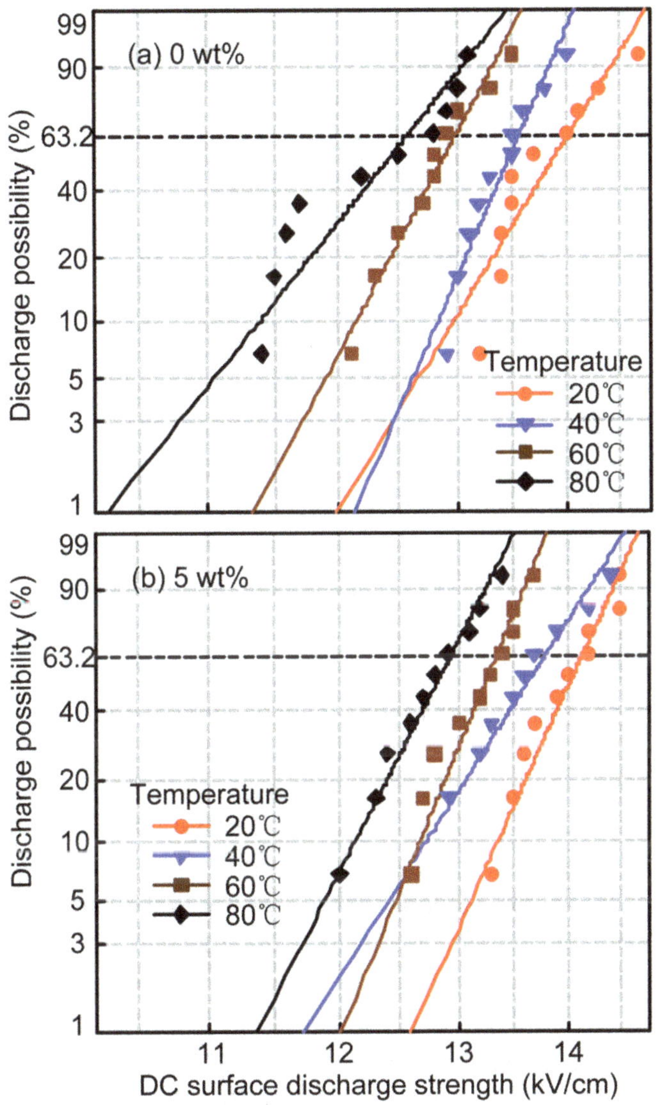

no nanoparticles were added had a surface discharge intensity at 14°C of 14.01 kV/cm. The electric field strength of the surface flashover at 80°C decreased by about 10.28% compared with 20°C. For the sample with the filler content

Figure 8. Relationship between the dc surface discharge strength and the ambient temperature

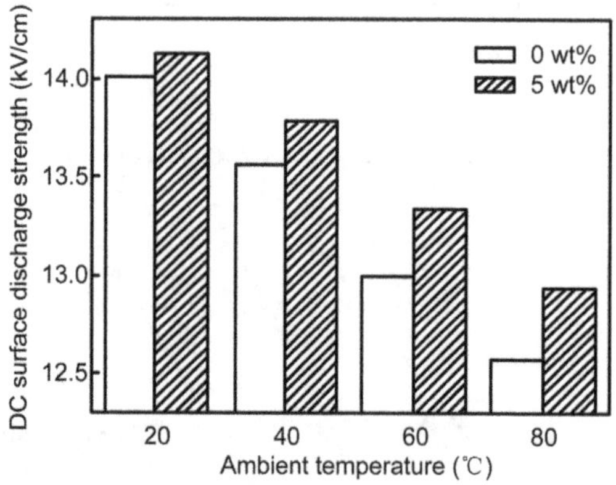

of 5 wt%, the electric field intensity of surface discharge decreased from 14.13 kV/cm at 20 °C to 12.94 kV/cm at 80°C.

According to the entropy increase theory, the charge is constantly moving. The thermal motion of the charge at the needle electrode is exacerbated by the increase in ambient temperature. The results of the surface potential decay experiment show that the surface charge is dissipated at a high ambient temperature, so the surface potential of the sample at the beginning of the dissipation process is low. On the one hand, the higher energy that carriers have is caused by higher ambient temperatures. Secondary electron emission avalanche is necessary for accelerating the development of surface discharge, and high energy electrons are liable to cause this phenomenon when the surface migrates. On the other hand, the carrier migration process is limited by traps around 0.85 eV. After the ambient temperature rises, the shallow traps near this energy level decrease, causing the effect of shallow traps to hinder charge migration. The presence of deep traps causes localized electric field concentration, because the charge trapped in deep traps is less likely to detach due to the need for more energy, which tends to cause local charge accumulation. The concentration of the local electric field can accelerate the electron emission process, thereby reducing the surface discharge intensity. Similar changes occurred in the flashover experiments of the original epoxy resin and epoxy resin nanocomposites.

Effect of Adding Nanoparticles

The effect of the addition of nanoparticles on the insulation properties of epoxy/aluminum nitride nanocomposites is reflected in the surface charge and discharge behavior of composites with different filler contents. Before the test was carried out at 80°C, the sample was first preheated on a heating platform, and the experiment was started when the temperature of the surface of the sample reached a steady state. Figure 9 shows the initial charge density of the sample surface and the surface charge decay rate during discharge calculated using the equation, with 60 s as the decay time.

At room temperature, the surface charge density of the sample to which the nano-aluminum nitride was added at the beginning of the dissipation process was higher than that of the epoxy resin sample to which no nanoparticles were added. Inaddition, the nanocomposite sample has a lower rate of surface charge density decay as reflected by the surface potential decay rate compared to the original epoxy resin sample. Among all the nanocomposites, the sample

Figure 9. The initial surface charge density and decay rate at (a) 20°C and (b) 80 °C

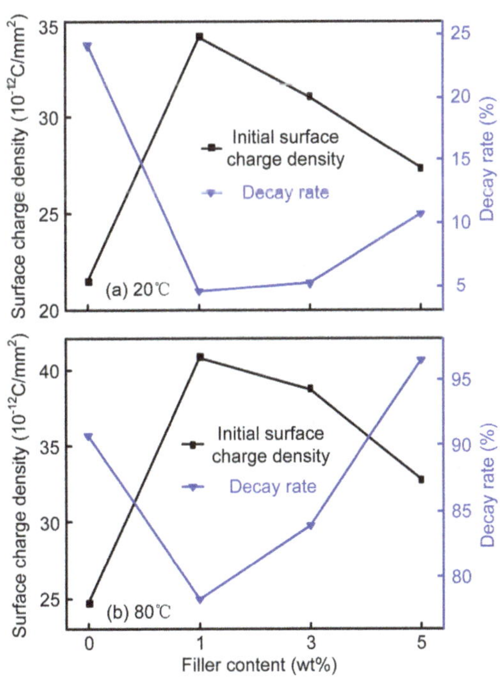

with a filling amount of 1 wt% had the highest surface charge density in the initial stage of dissipation, and the lowest rate of potential decay during the dissipation process. The surface charge density of the epoxy resin sample containing 1 wt% of alumina particles at the beginning of the dissipation phase was increased by about 66.7% compared to the original epoxy resin sample. As the amount of addition increases, the surface charge density at the beginning of the dissipation process gradually decreases, and the rate of potential decay during the dissipation process gradually increases. At 80°C, the density of the surface charge at the initial stage changes with the amount of nanoparticle added as in the case of 20°C. Under this condition, however, the surface charge density decay rate of the sample having a nanoparticle addition amount of 5 wt% was faster than that of the epoxy sample to which no nanoparticle was added.

The trap distribution of samples prepared by adding nanoparticles of different relative mass ratios is shown in Figure 10. At 20°C, the trap level and trap density of the epoxy/nano-AlN composites are higher than those of the samples without added nanoparticles, and the 1 wt% content of the sample has a higher trap energy level and the corresponding trap density. Both the number of traps and the trap energy decrease as the amount of nanoparticles added is further increased. The addition of nanoparticles has created a new trap for epoxy resins. At the same time, the agglomeration caused by the large addition of nanoparticles reduces the energy of the trap and the corresponding trap density.

The trap profile of the sample at 80°C shows two peaks, as described in Section 2.2 above. The first peak is the energy level density relationship of the shallow trap with relatively low energy, and the second peak is the energy level density relationship of the deep trap with relatively high energy. In order to more clearly explain the distribution of traps with different energy states in the sample, the curvesare divided into two parts representing the density distribution states of the relatively shallow trap and the relatively deep trap, respectively. For shallow traps, the sample with added nanoparticles showed a slightly higher trap energy level than the original epoxy sample. Among all the samples, the sample with an added amount of 1 wt% had the highest trap energy level of up to about 0.84 eV, while the deep trap energy of the epoxy sample without the added nanoparticles was about 0.81 eV, which was the trap level at room temperature. The distribution has the same trend as the amount of nanoparticle added. However, the shallow traps of the samples added in amounts of 3% and 5% by weight were lower than the shallow traps of the epoxy samples to which no nanoparticles were added. Among all the

Figure 10. The specimens' distribution of (a) traps at 20°C as well as (b) shallow traps and (c) deep traps at 80 °C

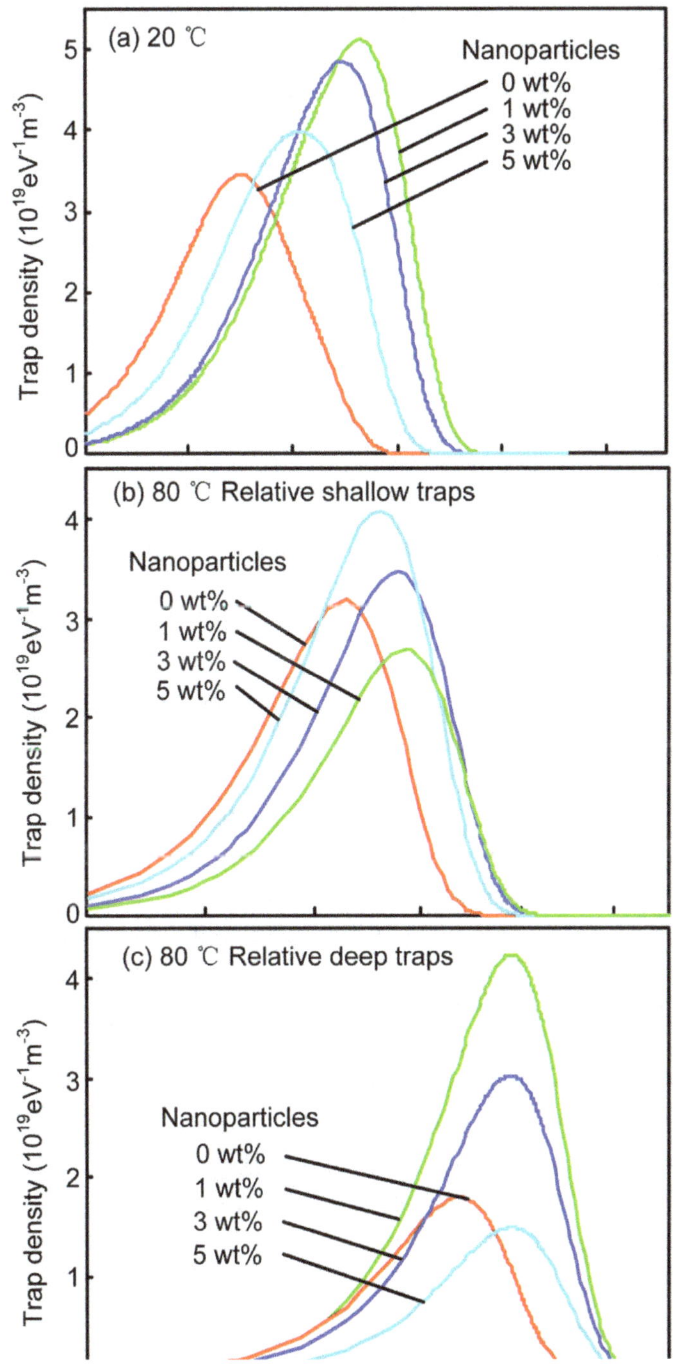

samples, the shallow traps of the samples added in an amount of 5 wt% were the most, and the shallow traps of the samples added in an amount of 1 wt% were the least. For traps with higher energy levels, the high energy level trap density of the sample with a loading of 1 wt% is greater. The deep trap density of the samples added in an amount of 1% and 3 wt% was higher than that of the epoxy sample to which no nanoparticles were added. The energy of deep traps in nanocomposites is also higher than the deep trap energy of epoxy samples without nanoparticles. However, the deep trap of the sample added in an amount of 5 wt% was less than the original epoxy sample.

Figure 11 shows the Weibull plot of the corresponding electric field strength when the sample flashes across the surface. Figure 12 shows the corresponding electric field strength at 60°C and 80°C for DC flashover. For an ambient temperature of 20°C, the electric field strength of the epoxy/nano-AlN composite material when flashing across the surface is higher than that of the epoxy resin without the addition of nanoparticles. The relationship between the intensity of the electric field and the amount of nanoparticle added and the trap level/density of the flashover are consistent with the trend of the amount of nanoparticle added. The epoxy sample without added nanoparticles had the lowest electric field strength required for creeping discharge, and the sample with the added amount of 1 wt% had the highest creep field strength. For the case where the ambient temperature is 80°C, the sample with an addition amount of 5 wt% has the highest creeping field strength. The creepage field strength of the sample added in an amount of 1 wt% was the lowest. The sample with a filler content of 3 wt% requires stronger electric field strength when the creeping discharge occurs than the sample with the filler content of 1 wt%, but the electric field strength is lower when flashing is compared with the pure epoxy sample without the added nanoparticles.

New energy traps are introduced into the epoxy resin by the addition of nanoparticles. The trap created by the addition of nanoparticles causes an increase in trap density in the sample, and the new trap energy is about 0.83 eV at 20°C. The trap at this time can be regarded as a shallow trap. The high-density shallow trapsexacerbate the charge trapping and detrapping process during the surface charge transfer of the insulating material sample. The sample with a mass fraction of 1 wt% exhibited the highest trap density and creeping field strength, and the experimental results were in accordance with the above theory.

Figure 11. Weibull plots of the surface discharge strength of the specimens at (a) 20°C and (b) 80 °C

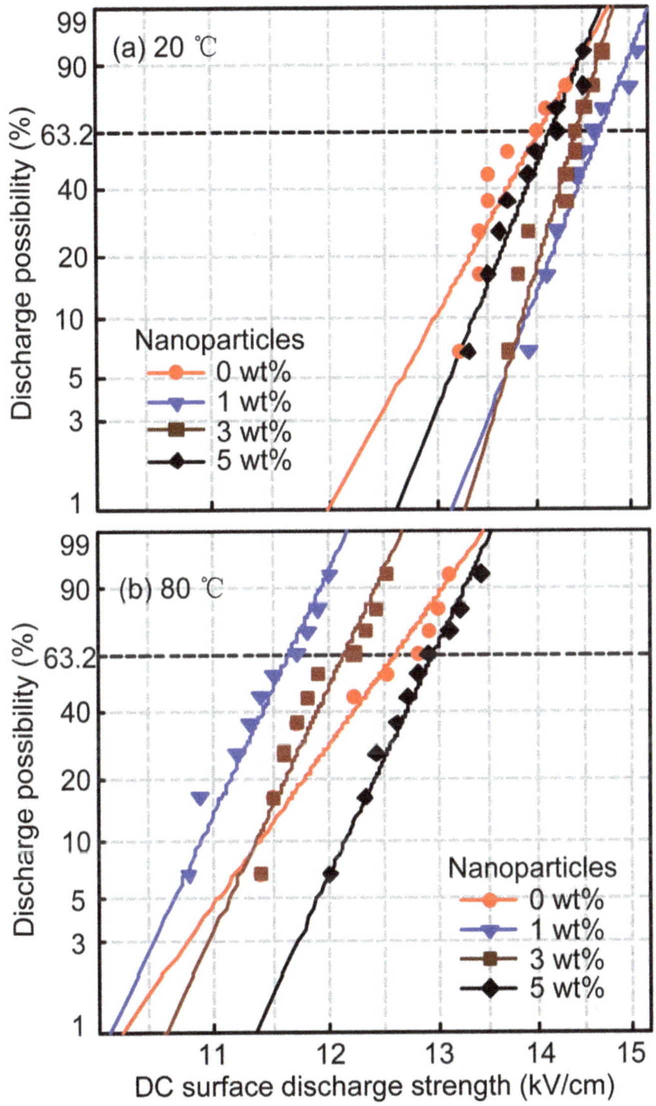

Figure 12. Relationship between the surface discharge strength and the filler content

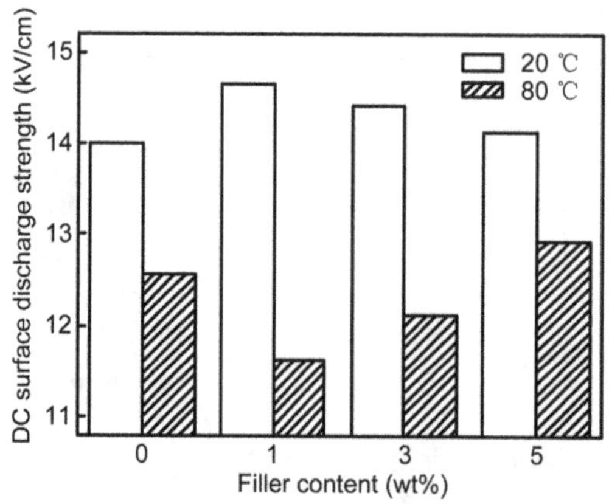

For the flashover phenomenon at 80°C, the electrons have higher energy at high ambient temperature. With the vigorous movement of high-energy electrons, the electric field strength required for the flashover of the insulating material decreases. When the ambient temperature is 80°C, the trap of the sample can be described as a relatively shallow trap with an energy level distribution around 0.83 eV and a relatively deep trap with an energy level around 0.9 eV. The trapping and detachment of charge is more likely to occur in relatively shallow traps, which will hinder the formation of electron avalanches. Conversely, the charge trapped by deep traps requires more energy to collapse and is less prone to collapse. The concentrated charge forms a local electric field, causing electric field distortion, accelerating the electron emission process, thereby weakening the surface discharge field strength. Samples with a nano-AlN particle loading of 5 wt% have more relatively shallow traps, while relatively deep traps have a lower density. At the same time, the sample with a nanoparticle addition amount of 5 wt% has higher electric field strength at 80°C than that of the epoxy sample to which no nanoparticle is added, and thus conforms to the above theory. Samples with filler mass fractions of 1% and 3% exhibited higher relative deep trap densities. The deeper the trap density, the more serious the local electric field concentration phenomenon will eventually reduce the electric field strength of the sample surface DC flashover. The results show that the electric field strength of the surface discharge in the sample with a filler content of 1 wt% at 80°C is the lowest.

IMPROVED CARRIER MOBILITY DEPENDENT SURFACE CHARGE AND FLASHOVER VOLTAGE OF POLYPROPYLENE FILM UNDER DC AND PULSE VOLTAGES

Surface Charge Accumulation and Decay with Different Voltages

In order to obtain and understand the characteristics of surface charge transport, carrier mobility needs to be calculated. After applying a DC voltage of ±4.5 kV to the sample with the surface modification time of 15, 30, 45, 60 min and the unmodified sample, the carrier mobility calculation results obtained by the surface potential decay experiment are shown in Figure 13. The results show that the surface modification has a significant effect on the carrier mobility of the sample after the application of positive and negative DC voltage. As the surface modification time prolonged, the carrier mobility increased from 0 to 45 minutes, and the carrier mobility decreased from 45 minutes to 60 minutes. This result is derived from the substitution of hydrogen on the surface of the surface-modified polypropylene film to form a dense fluoride on the surface of the film, which changes the lower surface energy level and higher trap distribution. However, when fluorine gas is applied to the surface of the sample for 60 min, excessive fluorination treatment will

Figure 13. Relationship between carrier mobility and modification time

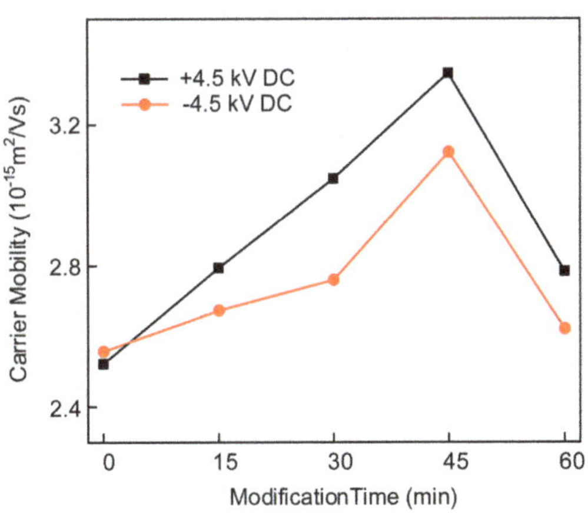

destroy the molecular chain structure on the surface of the film, forming a large number of chain scission and polar groups, resulting in an increase in deep traps with higher energy levels in the PP film. Charge transport is inhibited. Therefore, when the sample is fluorinated for 60 minutes, the rate of carrier migration decreases, and the mobility of carriers under positive DC voltage in the sample after surface modification is higher than that under negative DC voltage. rate. The high electronegativity of fluorine causes the surface of the film to absorb negative charges, weakening the transport process of negative charges (Du, 2018). The experimental results show that the surface modified 45 min sample has a good effect in improving the carrier mobility. Therefore, this section mainly selects the surface modified 45min sample and the original sample for comparison, and studies the surface charge dynamic behavior under the combined action of DC and pulse voltage.

To investigate the effect of surface fluorination on the carrier mobility of DC and pulsed composite voltages, the sample was first charged with a DC voltage and then a repetitive pulse voltage was applied to the sample. The DC voltage has an amplitude of 4.5 kV and is constant under all combinations. Figure 14 shows the carrier mobility at DC and pulse voltages of the same polarity. In Figure 14, a pulse voltage of 0 kV means that only a simple DC voltage is applied to the sample.

As can be seen from Figure 14(a), the migration rate of carriers in the modified sample is higher than that of the unmodified sample. In addition, the rate of carrier migration is also affected by the magnitude of the pulse voltage. As the amplitude of the pulse voltage increases, the carrier mobility of the original sample decreases first, and then the carrier migration rate increases with the further increase of the amplitude. In addition, the curve has a turning point at a pulse voltage of 6 kV. This phenomenon is mainly due to the close correlation between the migration rate of carriers and the electric field. After the application of a direct current voltage, charges accumulate on the surface of the sample, and surface charges establish an electric field between the surface of the sample and the ground electrode. When the amplitude of the pulse voltage is relatively small, the applied pulse voltage has less influence on the electric field between the two surfaces of the sample which has been established after the application of the direct current voltage. At low electric fields, the energy obtained by the charge is reduced, so that the charge is hard to collapse, and the rate at which the charge migrates from the surface of the sample to the ground electrode is slowed down. Therefore, the carrier mobility decreases as the pulse voltage increases. As the amplitude of the pulse voltage is further increased, the electric field between the two surfaces

Figure 14. Relationship between carrier mobility and pulse voltage

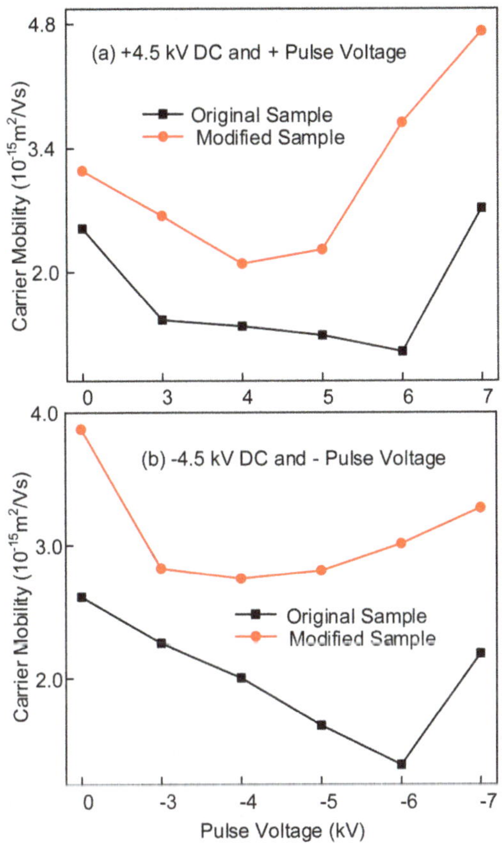

of the sample gradually changes to mainly depend on the pulse voltage. The electric field strength increases as the amplitude of the pulse voltage increases. The charge originally accumulated on the surface of the sample under a high electric field is more likely to be trapped, so the carrier migration rate is increased. The carrier mobility of the modified sample also decreases with the increase of the pulse voltage amplitude, and then increases with the further increase of the pulse voltage amplitude, but the turning point appears at the pulse voltage amplitude of 4 kV. At the same point, the pulse voltage amplitude of the inflection point corresponding to the pure PP sample is lower. In addition, the carrier migration rate of the modified sample after the corresponding pulse voltage at the turning point is higher than that of the original sample without surface treatment. For the combination of the negative DC and negative pulse voltages in Figure 14(b), the mobility tends to be the same as the pulse voltage amplitude, except that the turning point is usually smaller than the turning point of the positive polarity.

Surface Charge Accumulation

During charging, the tip corona discharge generates a charge, and some of the charge reaches and accumulates on the surface of the sample to form a surface charge. The thickness of the polypropylene film used in the experiment was 40 μm, and the surface charge was mainly concentrated on a thin layer having a depth of several micrometers below the surface of the sample. Therefore, it can be considered that the depth of the charge entering body is several micrometers. The instantaneous value of the surface potential was measured after the corona discharge, and the surface charge density was calculated as the initial surface charge density.

In order to study the dynamic behavior of the lower charge on the surface of the sample under the action of ±4.5kv DC voltage, the carrier mobility and initial surface charge density under different conditions are shown in Figure 15. The value of the vertical coordinate represents the absolute value of the initial surface charge density. The carrier transfer rate on the horizontal coordinates corresponds to the value in Figure 13 and is plotted in ascending order. The initial surface charge density decreases with increasing carrier mobility, indicating that higher carrier mobility can promote the transport of

Figure 15. Relationship between surface charge density and carrier mobility

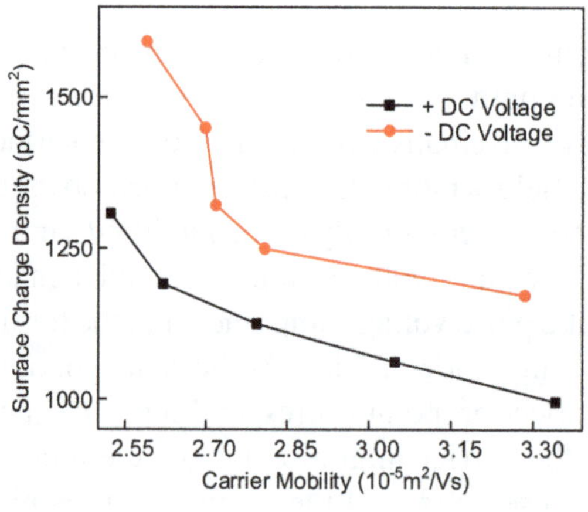

surface charge to the ground electrode. The study also found that the surface charge density at the beginning of the dissipation process at negative DC voltage is higher than the surface charge density at the initial stage of the positive DC voltage. The carrier mobility at a negative DC voltage is lower than the carrier mobility at a positive voltage, resulting in a higher surface charge density at the initial stage of the negative voltage. In addition, the needle plate electrode is more susceptible to corona discharge under the action of a negative electric field, generating more charge and accumulating on the surface of the sample.

Figure 16 shows the surface charge density at the beginning of the dissipation process after the pulse voltage of the same polarity is superimposed on the DC voltage. In Figure 16(a), when the positive polarity DC and positive polarity pulse voltages are applied to the sample one after the other, the surface charge density of the unmodified sample at the beginning of the dissipation process is first reduced after the application of the pulse voltage. the trend of. As the amplitude of the pulse voltage gradually increases to 6kV, the surface charge density at the beginning of the dissipation process changes from a gradual decrease to a gradual increase as the pulse voltage amplitude changes. When the applied pulse voltage amplitude is greater than 6 kV, the surface charge density at the beginning of the dissipation phase begins to increase. The results show that after the corona discharge process under the action of DC voltage, the surface of the film accumulates the charge generated by the discharge of the pin-plate electrode. When the pulse voltage is applied to the needle electrode for corona discharge, the same polarity charge will be Forced drive to the sample surface. The migration of carriers inside the sample and the repulsion between the pulse of the same polarity charge and the original charge accumulated by the DC voltage are two aspects that affect the surface charge density at the beginning of the dissipation process. The charge on the surface of the membrane migrates to the ground electrode under the action of the electric field within the membrane. In addition, the repulsion generated between the charge generated by the corona discharge and the charge accumulated on the surface of the sample caused by the DC voltage after the pulse voltage is applied to the needle electrode also suppresses the charge generated by the corona discharge of the pulse voltage from reaching the sample. surface. Although the carrier mobility decreases as the amplitude of the pulse voltage increases, the repulsion between the same type of charge reduces the charge reaching the surface of the sample, resulting in a decrease in surface charge density. However, when the pulse voltage is sufficiently high, under the action of a sufficiently strong pulsed electric field, more of

Figure 16. Relationship between surface charge density and combination of DC and pulse voltage with same polarity

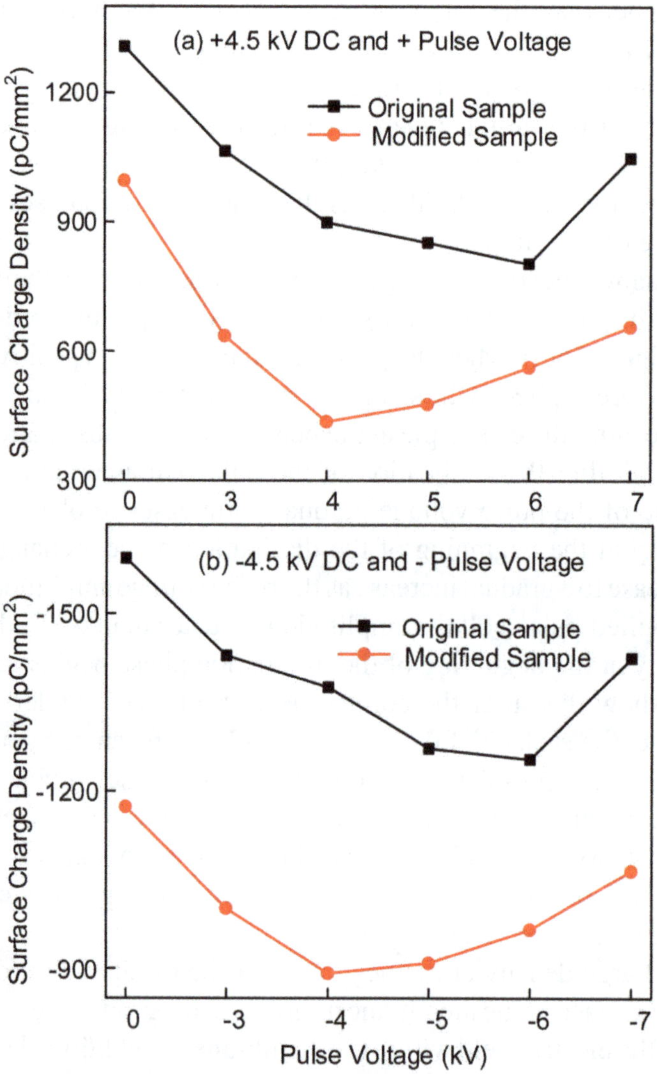

the same polarity charge will have enough energy to overcome the repulsive force of the electric field formed by the surface of the same polarity charge to reach and deposit on the surface of the sample. At this stage, the surface charge also rapidly migrates to the ground electrode due to the increase in carrier mobility. When the amount of charge newly injected into the sample by the pulse voltage exceeds the amount of charge dissipated inside the sample due to carrier migration, the surface charge density will increase.

The surface charge density of the sample after surface modification treatment has a similar trend with the unprocessed sample as the pulse voltage amplitude changes at the beginning of the charge dissipation process, but the corresponding surface charge density is lower, and the pulse voltage corresponding to the turning point The amplitude is 4kV. It can be inferred that the charge is easily dissipated after surface modification, and it is difficult for the charge to deposit on the surface of the sample. Therefore, the modified sample has less charge deposited on the surface of the original sample, and also causes a relatively smaller repulsive force between the charge generated by the pulse voltage and the charge accumulated on the surface. Therefore, a relatively small pulse voltage amplitude results in an increase in surface charge density. These factors lead to changes in the position of the turning point and the corresponding charge density shown by the experimental results. At the same time, it is also confirmed that the surface charge amount at the beginning of the dissipation process is correct with the pulse voltage amplitude under the action of the above composite voltage. For the composite voltage composed of the negative DC and negative pulses shown in Figure 16(b), the corresponding change law is similar except that the surface charge density value at the initial stage of the dissipation process is higher than the value under the positive polarity voltage. . In one aspect, the amount of corona discharge of the needle electrode at a negative voltage is greater than the amount of corona discharge at a positive voltage. On the other hand, the rate at which hole carriers and electrons migrate along the surface of the sample is different. The results show that when the charge injection rate and the dissipation rate reach equilibrium, the surface charge density at the beginning of the dissipation process is higher.

Figure 17 shows the charge density of the sample surface at the beginning of the dissipation process when the polarity of the DC voltage is opposite to the polarity of the pulse voltage. The curve in Figure 17 presents a single trend. In Figure 17(a), when the pulse voltage is 0 kV, it means that a simple DC voltage is applied to the electrode, and a positive charge is accumulated on the surface of the sample. When a negative pulse voltage is applied, the negative charge is induced to fall to the surface of the sample. The mutual attraction between charges having opposite polarities makes the negative charge generated by the corona discharge of the pulse voltage more easily reach the surface of the sample. Therefore, a relatively small pulse voltage

Figure 17. Relationship between surface charge density and combination of DC and pulse voltage with different polarity

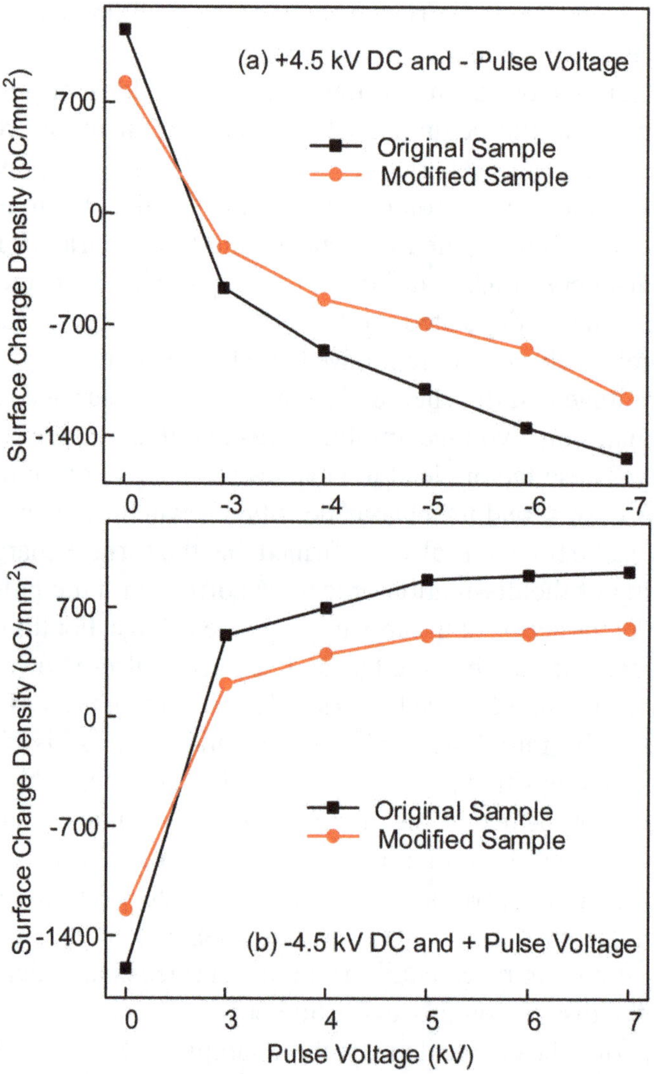

amplitude produces an opposite polarity charge that is sufficient to neutralize the previously deposited charge. In addition, some residual charge will fall on the surface of the sample. As the pulse voltage increases, the surface charge density at the beginning of the dissipation process also increases accordingly.

At the same voltage, the original sample surface accumulates a higher charge density at the beginning of the dissipation process than the modified sample. A similar trend can be observed in Figure 17(b) when the sample is in a composite of a negative DC voltage and a positive pulse voltage.

Flashover Voltage

Flashing on the surface causes the dielectric of the film capacitor to fail. Figure 18 shows the flashover voltage of the sample after surface modification at different DC and pulsed composite voltages. Regardless of the DC and pulse polarity, when the surface fluorination modification time increases from 0 to 45 min, the flashover voltage will increase, and when the modification time continues to increase, the flashover voltage will decay again. The trend and carrier mobility are similar to the trend of surface modification time. The results show that the charge accumulated on the surface of the sample has a certain influence on the flashover resistance of the material. It is known from the above initial surface charge density that the surface modification causes a decrease in the surface charge, so that the local concentration of the electric field caused by the electric charge is alleviated, thereby suppressing the generation of the surface discharge. The results show that the flashover voltage under the pulse voltage is higher than the flashover voltage under the DC voltage regardless of the polarity, and the flashover voltage under the negative voltage condition is higher. The c-f bond in the fluorinated layer of the dielectric film has strong chemical stability, increases the inertia of the surface of the dielectric film, and can suppress the acceleration of the negative electric charge under the action of the electric field when migrating along the surface of the film, thereby suppressing the occurrence of electron avalanche. The results show that the surface fluorination modification treatment improves the flashover voltage of the sample, and the effect at the negative voltage is particularly obvious. When the surface fluorination treatment time is longer than 45 min, the amount of charge accumulated on the surface of the sample increases, resulting in severe local electric field concentration. Subsequent strong electric field inhomogeneity introduces micro-discharge, which generates a large amount of primary electrons for secondary electron emission, accelerates the formation of electron avalanche, and finally flashes along the surface of the film.

Figure 18. Relationship between flashover voltage and modification time under DC and pulse voltage respectively

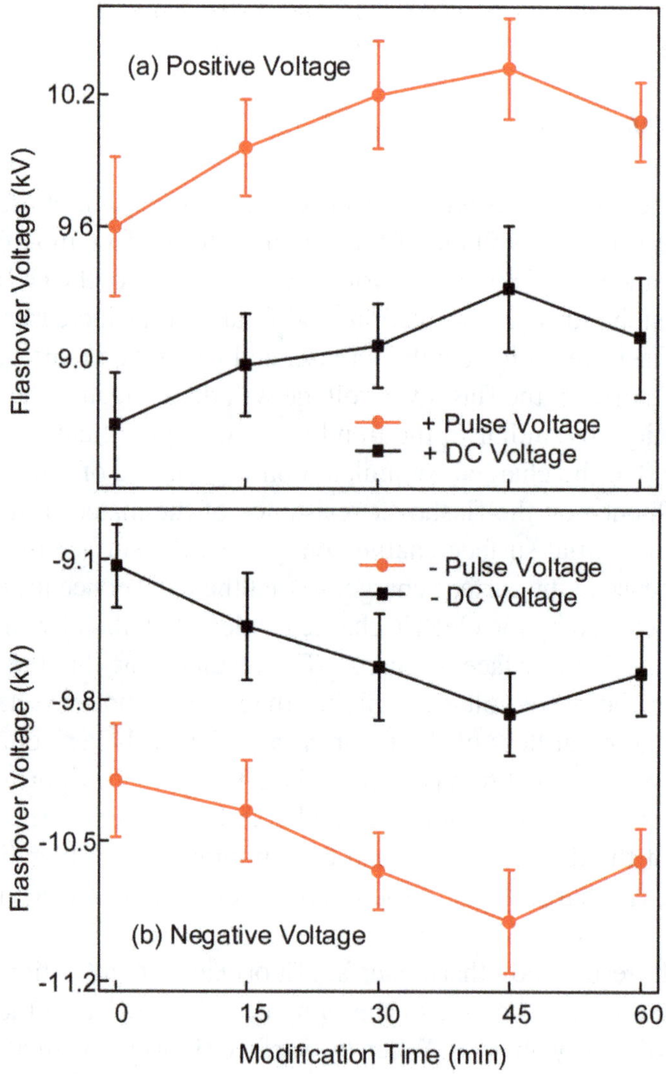

Figure 19 shows the effect of surface modification treatment time on the flashover voltage of the DC voltage superimposed pulse voltage. The flashover voltage in vertical coordinates shows the sum of the DC voltage and the pulse voltage amplitude. It can be seen that the flashover voltage of the material

Figure 19. Relationship between flashover voltage and modification time under superimposed voltage of DC and pulse voltage

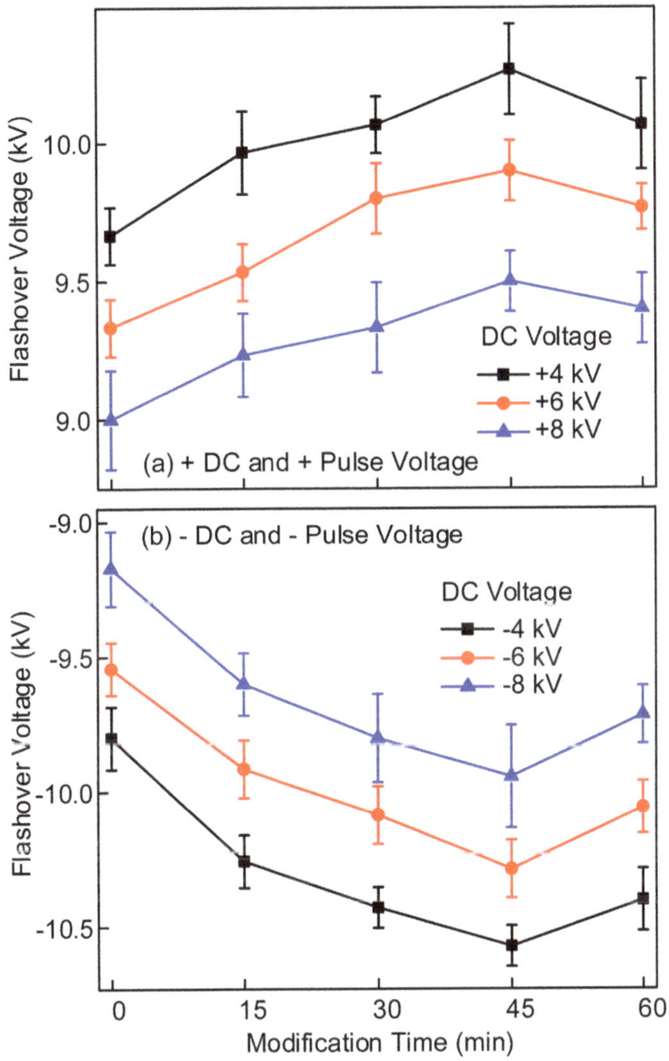

under the composite voltage has the same tendency as the flashover voltage under the condition of pure DC voltage or pure pulse voltage. The results show that the surface fluorination modification treatment can improve the flashover resistance of the insulating material under the composite voltage.

When the DC voltage is increased from +4 to +8kv, the flashover voltage values of both polarities are correspondingly reduced. The results show that the charge in the DC voltage electric field can get more energy than the charge in the pulse voltage electric field. Under the same modification time conditions, at a higher DC voltage, only a small pulse voltage is required to cause flashover. In addition, when a DC voltage of the same magnitude is applied to the sample, in the case where the surface modification time is the same, when the superimposed voltage polarity is negative, the corresponding voltage at the time of flashover is higher.

EFFECTS OF GRAPHENE NANOPLATELETS ON SPACE CHARGE AND BREAKDOWN STRENGTH OF PP/ULDPE BLENDS FOR HVDC CABLE INSULATION

Specimens Characteristics

The sample was first quenched with liquid nitrogen, and the distribution of graphene in the PP-based blend and the phase structure of ULDPE were observed by observing the fracture surface using a scanning electron microscope (SEM). As shown in Figure 20a, the distribution of ULDPE in the blend is described from the perspective of the PP/ULDPE interface, and it can be observed that the blend is formed by coating a small ULDPE droplet having an irregular shape with a PP matrix. Taking PP/ULDPE/graphene nanocomposites as an example, the morphology is shown in Figure 20c, where the ratios of PP, ULDPE and graphene are 85 wt%, 15 wt% and 0.01 wt%, respectively. The medium graphene particles are uniformly dispersed with almost no agglomeration. As can be seen from Figure 20d, the graphene in the sample pu15g5 showed partial agglomeration in the mixture. Furthermore, in Figure 20d, the overlapping portion that is most likely to form a local low resistance path is marked with a red dashed line.

The mechanical properties of insulating materials are one of the important parameters to consider when developing cable insulation. The parameters used to characterize the toughness and stiffness of the polymer include elongation at break and tensile strength. In order to find a suitable PP/ULDPE blend formulation with good mechanical properties, tensile tests were performed on pure PP and PP/ULDPE blend samples. Figure 21 shows the dynamic stress-strain curve for a PP/ULDPE blend where strain is an independent

Figure 20. SEM images of sample fracture surface: (a) PU15, (b) PU15G0.5, (c) PU15G1, (d) PU15G5

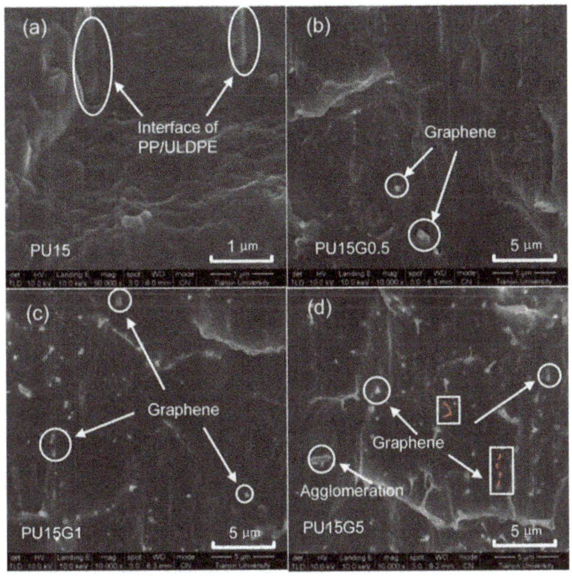

Figure 21. The stress-strain curves of PP/ULDPE blends

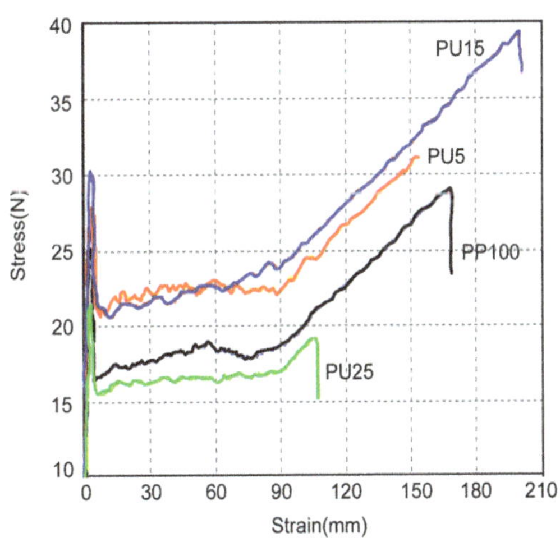

variable and stress is a dependent variable. The results show that the four samples exhibited similar yield behavior during the tensile test.

Figure 22 shows the results of the elongation at break test and the tensile strength test. In this section, the rigid matrix is polypropylene and the dispersed ULDPE acts as a rubber domain in the blend. As the ULDPE content increases, the tensile strength of the PP/ULDPE blend first increases and then decreases. Too low a content (5 wt%) and an excessively high content (25 wt%) of ULDPE will result in a decrease in the elongation at break of the PP/ULDPE blend compared to pure PP. When the ULDPE content reaches 15 wt%, the elongation at break and tensile strength of the blend reach a maximum; in cable insulation applications, the blend of this formulation has a better mechanical reinforcement than pure PP.

The size and dispersibility of the particles in the blend are important factors influencing the best mechanical properties of the blend. The crystalline phase (mostly spherulites) and the amorphous phase coexist in PP, so PP is a semi-crystalline polymer. The splitting of large spherulites into small dies can enhance the impact resistance of PP. In order to break the large spherulites of PP into smaller spherulites, ULDPE can be distributed in PP phase by mixing ULDPE particles with PP matrix. between. Low levels (5 wt%) of ULDPE are not sufficient as rigid modifiers, while high levels (25 wt%) of ULDPE have an effect on the crystallization of PP, resulting in mechanical properties that are worse than unpure pure PP. However, in the same blending mode, this uniform short-chain branching distribution characteristic of the 15 wt% content of ULDPE can make the interfacial coupling between the PP matrices stronger, thereby improving the tensile strength and impact toughness of the PP. Therefore, 15 wt% of ULDPE was first added to the PP to enhance the flexibility and tensile strength of the blend, and then PU15 was blended with graphene of different mass fractions to further conduct electrical tests on the blend.

Figure 23 shows the DC conductivity of the different samples. The conductivity of the PU15 blend is worse than that of the PP without the addition of ULDPE. This is because the interface formed by introducing ULDPE into the PP can capture charge carriers, thereby limiting the carrier mobility, resulting in a total of PP/ULDPE. The DC conductivity of the mixture is low. The low addition of graphene nanosheets with large specific surface area (eg 0.005 wt%) makes it easier to capture a large number of interaction zones at the micro interface formed between graphene and polymer than PP100 and PP15 to capture charge carriers. Therefore, the addition of nano-sized graphene limits the transport of charge carriers, making the nanocomposites

Figure 22. Tensile characteristics of PP/ULDPE blends

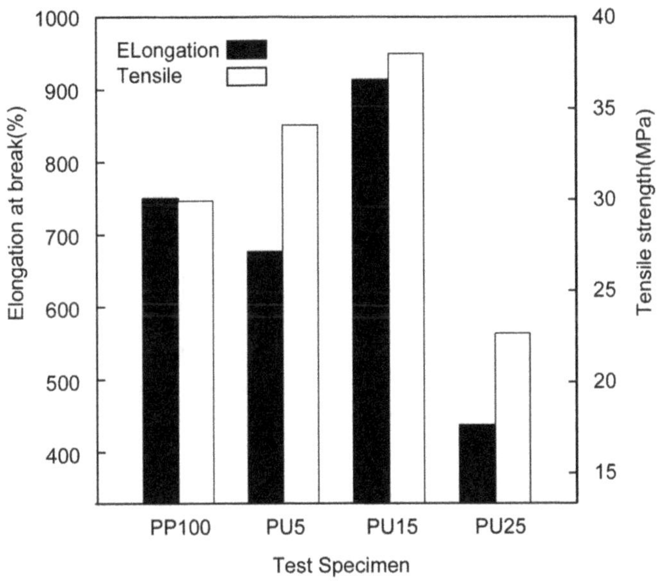

have lower DC conductivity than PP and PP/ULDPE blends. As shown in Figure 23, the PP/ULDPE blend having a graphene content of 0.01 wt% has a much lower DC conductivity than other composite materials. At the same time, when the graphene nanosheet content is further increased to 0.05 wt%, the direct current conductivity is greatly improved. When the graphene content is sufficiently high, the interaction zone between the polymer and the filler may overlap each other, providing more opportunities for forming a low-resistance path for electron and hole migration, accelerating the overlap of charge carriers. Transport of the agglomerated portion (rather than transport in the whole), resulting in an increase in DC conductance.

4.2. Space Charge Behaviors

In the design of high-voltage DC cable insulation, space charge behavior is one of the main problems, so it is necessary to pay attention to the space charge problem under DC when designing parameters. Figure 24 shows the space charge distribution in pure PP, PU15, PP/ULDPE/graphene nanocomposites after energization for 3600 s at a 60 kV/mm DC electric field and the

Figure 23. DC conductivity of different specimens under 20 kV/mm

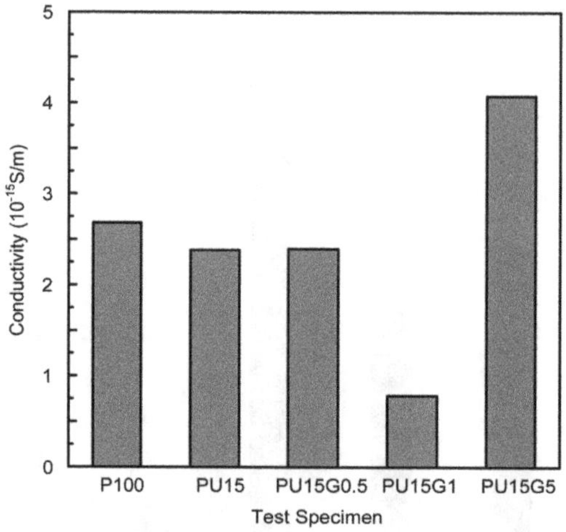

corresponding electric field distribution in the sample. Figure 25 shows the dynamic distribution of space charge in pure PP and PP/ULDPE/graphene nanocomposites that were not added during depolarization after the end of the polarization process.

To better compare the time dependence of the charge amount, calculate the total absolute charge for each sample according to the following function:

$$Q(t) = \int_0^L |\rho(x,t)| S dx \tag{4-1}$$

where $\rho(x,t)$ is the charge density in depth x (distance from the cathode) at poling or depolarization time t, S is the electrode area, L is the thickness of the specimen. The relationship between total charge and the poling time, the depolarization time are shown in Figure 26.

Figure 24a depicts the dynamic distribution of space charge in pure PP during polarization. Over time, only a small amount of impurity is present near the interface where the electrode is in contact with the sample. Since the injection of charge into the PP matrix requires overcoming a very high electric field threshold, as shown in Figure 26a, the total charge charge injected into the polypropylene during the polarization process under the action of a high voltage DC electric field increases slowly, at this time inside the substrate. The accumulated charge is mainly derived from some polarized or weakly

Figure 24. Space charge polarization distribution in PP and PP/ULDPE/graphene nanocomposites under 60 kV/mm and the electric field distribution after 3600 s

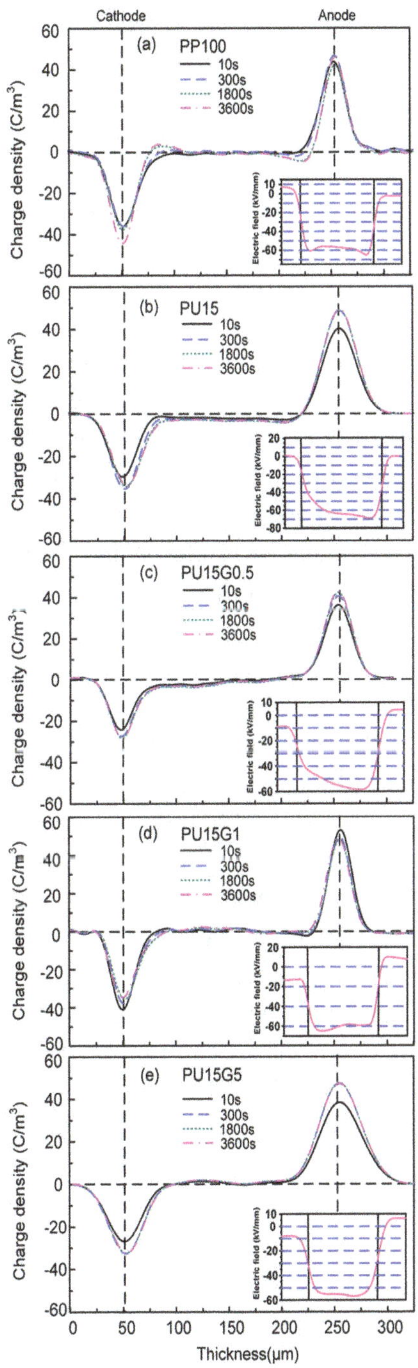

bound ions. The results of the electric field distribution show that the field distortion of pure polypropylene occurs only near the interface between the electrode and the sample, which is the same as the area where the space charge accumulates. Taking 3s as an example, it can be seen from Figure 25 that the charge injected into the sample during the polarization at the beginning of the depolarization process cannot be instantaneously released, and a small amount of charge may be concentrated in the vicinity of the contact faces of the sample and the electrode. As can be seen from Figures 25a and 26a, the rate of charge carrier migration is low, resulting in a slow decay of space charge accumulated in pure PP after the end of the polarization process during volt-off.

As shown in Figures 24b and 26a, a large amount of space charge is injected from the electrode and confined near the interface of the PU 15 and the electrode. This is due to the introduction of ULDPE, which leads to a large number of interfaces in pure PP. There are many trapping electron traps on the interface, which leads to the accumulation of space charge at the interface. Therefore, local electric field occurs in the body of PU15 and near the interface between PU15 and the electrode. concentrated. As shown in Figure 26b, the space charge in PU15 dissipates rapidly at 30 s, and then the dissipation rate is slowed down due to the limitation of deep traps.

By performing a synergistic analysis of Figures 24c, 24d, and 24e with Figure 24b, the total charge shown in Figure 26a can be verified. The introduction of graphene improves the space charge dynamics in PP/ULDPE blends. As shown in Figures 24c and 26a, the space charge accumulated in the nanocomposite prepared by adding graphene nanosheets is less than that in PU15. Figures 24 and 26a show that the charge injected from the electrode into the PU15G1 sample is much less. The electric field distribution inside the sample at 3600 s can be calculated from the accumulation of space charge. The electric field strength near the cathode portion of PU15G1 is up to 64.5 kV/mm, but lower than the maximum electric field strength at the corresponding position in PU15 (69.8 kV/mm). Graphene and PP/ULDPE blended to form a large number of micro-interfaces, and the interaction zone on the interface inhibited the transport of charge carriers, thus affecting the dynamic behavior of space charge. As the content of graphene nanosheets increases, the interaction zone in the PP/ULDPE blend increases. On this basis, the space charge distribution of PU15, PU15G0.5 and PU15G1 was observed. According to the results in Figures 25 and 26b, it can be seen that doped graphene nanosheets can also limit the mobility of charge carriers. Over time, the accumulated space charge dissipated slowly in pu15g1, and a large amount of space charge was still confined inside the sample after 600 seconds.

At the same time, as shown in Figures 24 and 26, the results show that the total amount of space charge accumulated in PU15G5 after the end of the polarization process is much less than that of PU15, and only a small amount of space charge is concentrated in the sample, between the upper and lower surfaces. In addition, the space charge in Figures 25e and 26b dissipates very quickly, with almost no charge remaining in the sample after 600 s. However, the reason for the low amount of space charge accumulation and the fast dissipation rate is slightly different from the previous explanation, which is mainly due to the fact that the charge carriers are transported in the local high conductance path (not the integral of the insulating matrix formed in the PP/ULDPE blend). This result can be further confirmed in the next section, that is, PU15G5 has a lower trap level and a smaller trap density, which results in a composite material having a large graphene nanosheet addition amount having lower breakdown strength. Since the total amount of space charge accumulated in the graphene nanocomposite is smaller than that of the simple PU15 to which the nanosheet is not added, the electric field is hardly distorted.

Trap Level Distribution and Breakdown Strength

The method of isothermal discharge current (IDC) can be used to test and calculate the trap distribution characteristics of different graphene content samples. Figure 27 shows the isothermal discharge current (i) of a PP/ULDPE composite with different graphene nanofiller contents. The dependent variable is the depolarization time (t). At the initial stage of discharge, the isothermal discharge current value of pure PP is the highest. PU15G1 and PU15G5 are the two largest and smallest discharge currents in the composite materials used in the experiments, respectively. The charge in the shallow trap first undergoes a desorption process that is faster than the charge in the deep trap. Therefore, the discharge current may be higher in the early stage. Subsequently, higher-level traps begin to function during charge release (Du, 2018). Thus, it is observed that the discharge current value gradually decreases with the depolarization time, and finally reaches a steady state.

Figure 25. Space charge depolarization distribution in PP and PP/ULDPE/graphene nanocomposites after the polarization of 3600 s under 60 kV/mm

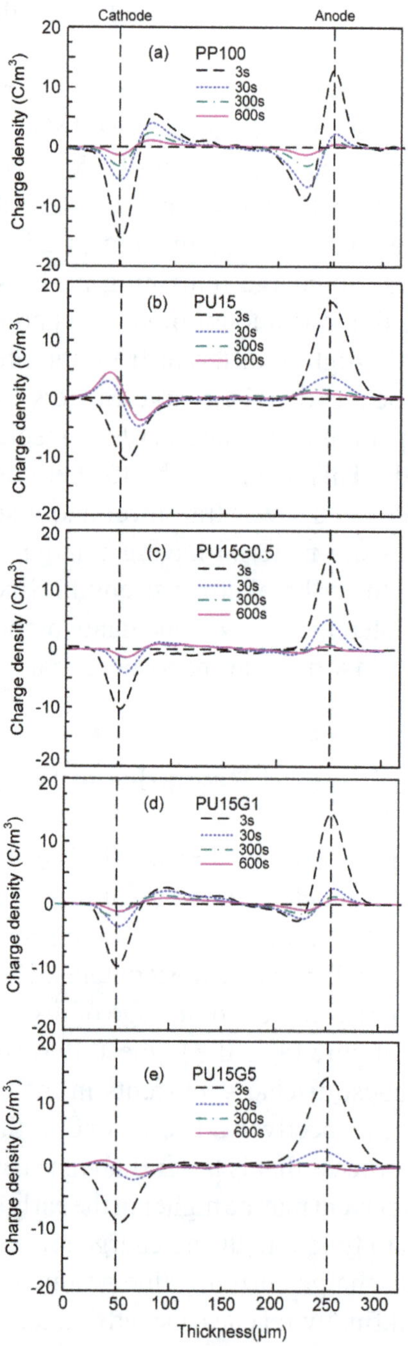

Figure 26. Relationship between total charge and (a) the poling time, (b) the depolarization time

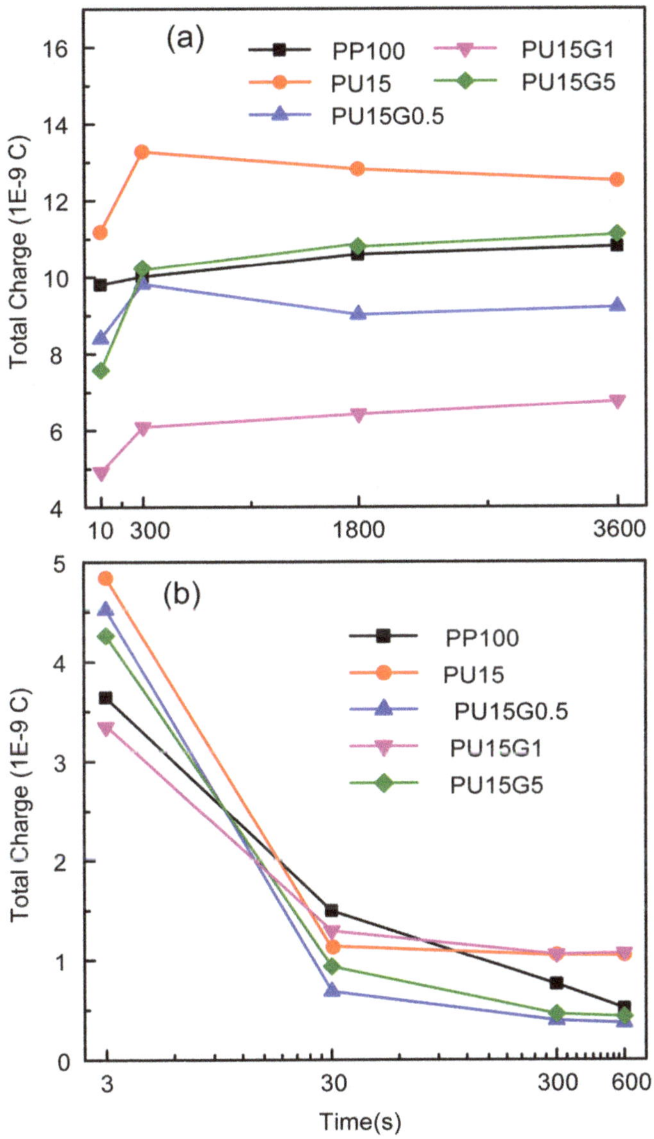

As shown in Figure 28, the trap level of the sample ranges from 0.76 to 1.1 eV. The results show that the PP and PP/ULDPE/graphene nanocomposites with different filler contents have two trap level density peaks. The peak value of the first trap level is about 0.96 eV, and the peak value of the second trap level is about 1.02 eV. The results show that PP has the highest trap level density, about 0.96 eV, and has more shallow traps than other samples. The trap level density of PU15 is lower than pure PP in shallow trap level and deep trap level, because the smaller size spherulites in PP/ULDPE blends are compared to the large spherulites in the original pure PP. Said that the trap on the edge of the spherulites is reduced. As shown in Figure 27, the trap level distribution is consistent with the isothermal discharge process.

As the content of graphene nanofiller in PU15 increasing from 0% to 0.01 wt%, the density of deep traps increases. PU15g1 achieved the highest trap level density at approximately 1.02 eV with a trap depth greater than other samples. This is attributed to the large specific surface area of the graphene nanosheets having an atomic layer thickness, which can form deeper traps and

Figure 27. Relationship between the IDC and the time

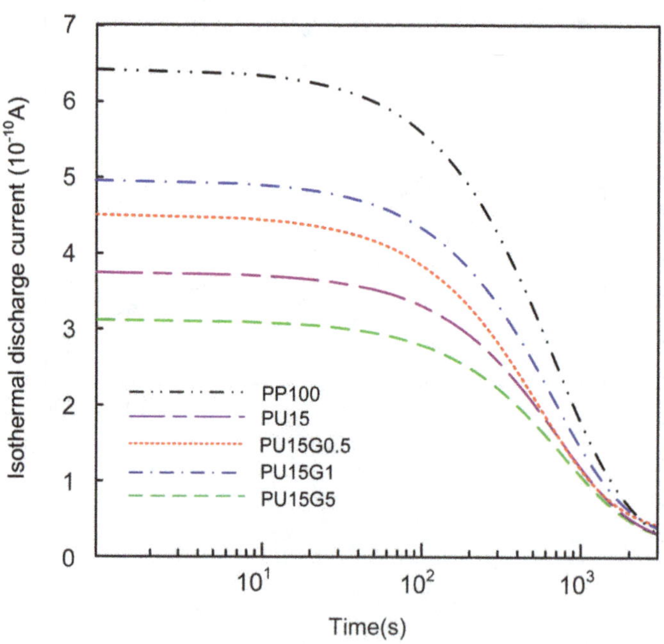

capture many electrons or holes. The movement of these trapped traps will be limited. In addition, the charge trapped on the surface of the sample forms a local electric field that weakens the electric field generated by the electrode, thereby suppressing charge injection at the electrode and suppressing the formation of space charge in the sample body. When the nanofiller content of graphene is increased to 0.05 wt%, more graphene nanosheets distributed in the matrix are in contact with each other, which provides a low resistance path for local transport of electrons and holes, thereby reducing the pair capture of charge.

The Weibull distribution of the DC breakdown strength of PP/ULDPE/graphene nanocomposites is shown in Figure 29, and the breakdown field strength corresponding to the 63.2% probability under each condition is indicated. The results show that the PF DC breakdown strength decreases slightly after adding 15% ULDPE. As shown, when the graphene content is low, the DC breakdown strength of the PP/ULDPE (pu15) blend increases first, and then a slightly higher addition reduces the breakdown strength of the blend. In contrast, as the nanofiller content increased from 0.005 wt% to 0.01 wt%, the breakdown strength of PU15g0.5 and PU15g1 increased. The results show that the interface trap has a great influence on the breakdown strength of the nanocomposite. Graphene nanoplates in PP/ULDPE blends can introduce a large number of deep traps, limiting the transport of carriers.

Figure 28. Relationship between the trap level and the trap level density

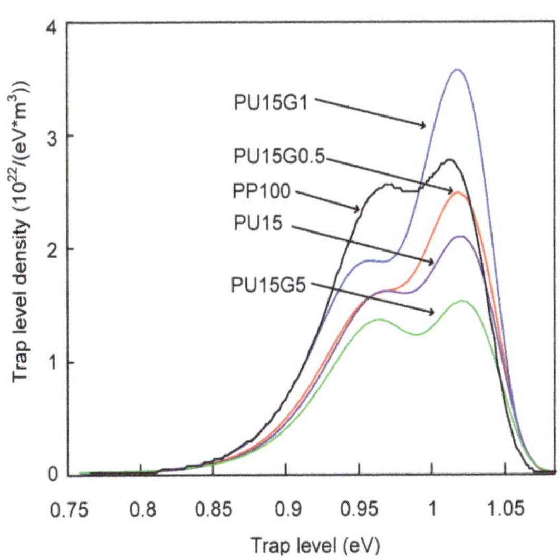

Figure 29. Weibull distribution of dc breakdown strength

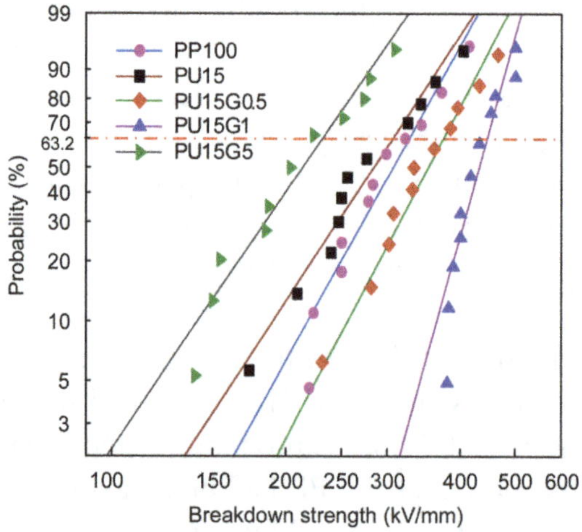

The increase in trap energy makes carrier migration need to overcome higher barriers, thereby increasing polymer chain breaks and improving the breakdown strength of the material. When the graphene content reaches 0.05% by weight, the DC breakdown strength of PU15g5 is remarkably lowered. At relatively high concentrations (as in the 0.05 wt% of this study) there is more contact between the graphene nanosheets in the matrix, which results in localized transport of charge carriers in the PP/ULDPE blend. Some high conductance paths cause a significant decrease in DC breakdown strength.

CONCLUSION

The carrier mobility and surface charge accumulation behavior of polypropylene films are affected by surface modification time, voltage type and polarity. The results show that after surface modification, the surface charge accumulation of the polypropylene film is inhibited and the flashover voltage is increased. Improving the insulation properties of polypropylene films is important for capacitors. The mechanical properties of PP/ULDPE blends are closely related to the ULDPE content. Too low or too high content will result in poor mechanical properties. The space charge behavior and electric field distortion of PP/ULDPE blend (pu15) are more serious than PP. The introduction of low content graphene nanosheets inhibits the space charge accumulation of PU15 blend. The results show that the proper ratio of PP/ULDPE/graphene nanocomposites has a good application prospect in high voltage DC cable insulation. The higher the ambient temperature is, the deeper the trap level and the smaller the trap density are. The charge gets a higher energy at higher ambient temperatures. At room temperature, nanoparticles result in deeper trap levels and higher trap densities. At higher ambient temperatures, samples filled with more nanoparticles showed higher shallow trap densities. The surface discharge intensity depends on the migration and accumulation of charge. Ambient temperature and trap distribution have an effect on surface discharge behavior. The results of the study help to find the right formulation to increase the flashover voltage of the epoxy resin at different operating temperatures.

REFERENCES

Du, B. X., Hou, Z. H., Li, J., & Li, Z. L. (2018). Effect of graphene nanoplatelets on space charge and breakdown strength of PP/ULDPE blends for hvdc cable insulation. *IEEE Transactions on Dielectrics and Electrical Insulation*, *25*(6), 2405–2412. doi:10.1109/TDEI.2018.007271

Du, B. X., Kong, X. X., Cui, B., & Li, J. (2017). Improved ampacity of buried hvdc cable with high thermal conductivity ldpe/bn insulation. *IEEE Transactions on Dielectrics and Electrical Insulation*, *24*(5), 2667–2676. doi:10.1109/TDEI.2017.006452

Du, B. X., & Li, J. (2016). Effects of ambient temperature on surface charge and flashover of heat-shrinkable polymer under polarity reversal voltage. *IEEE Transactions on Dielectrics and Electrical Insulation*, *23*(2), 1190–1197. doi:10.1109/TDEI.2015.005182

Du, B. X., Wang, M. Y., Jin, L., & Xing, Y. Q. (2018). Temperature dependent surface charge and discharge behavior of epoxy/ain nanocomposites. *IEEE Transactions on Dielectrics and Electrical Insulation*, *25*(4), 1300–1307. doi:10.1109/TDEI.2018.007019

Du, B. X., & Xiao, M. (2013). Thermal accumulation and tracking failure process of bn-filler epoxy-matrix composite. *IEEE Transactions on Dielectrics and Electrical Insulation*, *20*(6), 2270–2276. doi:10.1109/TDEI.2013.6678879

Du, B. X., Xu, H., & Li, J. (2017). Effects of mechanical stretching on space charge behaviors of PP/poe blend for hvdc cables. *IEEE Transactions on Dielectrics and Electrical Insulation*, *24*(3), 1438–1445. doi:10.1109/TDEI.2017.006116

Du, B. X., Xu, R. R., Jin, L., & Li, Z. L. (2018). Improved carrier mobility dependent surface charge am flashover voltage of polypropylene film under dc and Pulse voltages. *IEEE Transactions on Dielectrics and Electrical Insulation*, *25*(3), 1014–1021. doi:10.1109/TDEI.2018.006845

Du, B. X., Zhu, W. B., Li, J., Xing, Y. Q., & Huang, P. H. (2017). Temperature-dependent surface charge behavior of polypropylene film under dc and Pulse voltages. *IEEE Transactions on Dielectrics and Electrical Insulation*, *24*(2), 774–783. doi:10.1109/TDEI.2017.006290

Flourentzou, N., Agelidis, V. G., & Demetriades, G. D. (2009). Vsc-based hvdc power transmission systems: An overview. *IEEE Transactions on Power Electronics*, *24*(3), 592–602. doi:10.1109/TPEL.2008.2008441

Grafton, M. A., Fothergill, J. C., Dissado, L. A., Shirley-Elgood, J. R. R., Stevens, G. C., & Thomas, J. L. (2001). Controlling flashover between electrode segments in DC power capacitors. *IEEE International Conference on Solid Dielectrics*. 10.1109/ICSD.2001.955623

Green, C., Vaughan, A., Stevens, G., Pye, A., Sutton, S., Geussens, T., & Fairhurst, M. (2015). Thermoplastic cable insulation comprising a blend of isotactic polypropylene and a propylene-ethylene copolymer. *IEEE Transactions on Dielectrics and Electrical Insulation*, *22*(2), 639–648. doi:10.1109/TDEI.2015.7076758

Huang, X., Fan, Y., Zhang, J., & Jiang, P. (2017). Polypropylene based thermoplastic polymers for potential recyclable hvdc cable insulation applications. *IEEE Transactions on Dielectrics and Electrical Insulation*, *24*(3), 1446–1456. doi:10.1109/TDEI.2017.006230

Khare, A. R., Westphal, S. P., Ling, M. T. K., Qin, C., & Woo, L. (2000). Thermal and dynamic mechanical analysis on metallocene ULDPE/PP blends to optimize impact properties. *Thermochimica Acta*, *357*(00), 155–160. doi:10.1016/S0040-6031(00)00384-1

Kong, M. G., & Lee, Y. P. (2001). Impact of surface discharge plasmas on performance of a metallized film capacitor. *Journal of Applied Physics*, *90*(6), 3069–3078. doi:10.1063/1.1389072

Kumara, S., Alam, S., Hoque, I. R., Serdyuk, Y. V., & Gubanski, S. M. (2012). Dc flashover characteristics of a polymeric insulator in presence of surface charges. *IEEE Transactions on Dielectrics and Electrical Insulation*, *19*(3), 1084–1090. doi:10.1109/TDEI.2012.6215116

Li, S., Huang, Q., Sun, J., Zhang, T., & Li, J. (2010). Effect of traps on surface flashover of xlpe in vacuum. *IEEE Transactions on Dielectrics and Electrical Insulation*, *17*(3), 964–970. doi:10.1109/TDEI.2010.5492273

Li, S., Zhao, N., Nie, Y., Wang, X., Chen, G., & Teyssedre, G. (2015). Space charge characteristics of ldpe nanocomposite/ldpe insulation system. *IEEE Transactions on Dielectrics and Electrical Insulation*, *22*(1), 92–100. doi:10.1109/TDEI.2014.004524

Montanari, G. (2011). Bringing an insulation to failure: the role of space charge. *IEEE Transactions on Dielectrics and Electrical Insulation*, *18*(2).

Shao, T., Liu, F., Hai, B., Ma, Y., Wang, R., & Ren, C. (2017). Surface modification of epoxy using an atmospheric pressure dielectric barrier discharge to accelerate surface charge dissipation. *IEEE Transactions on Dielectrics and Electrical Insulation*, *24*(3), 1557–1565. doi:10.1109/TDEI.2017.006321

Tanaka, T. (2005). Dielectric nanocomposites with insulating properties. *IEEE Transactions on Dielectrics and Electrical Insulation*, *12*(5), 914–928. doi:10.1109/TDEI.2005.1522186

Tsunoda, M., Tsuchiya, Y., Hashimoto, T., & Takahashi, M. (2000). Morphological characteristics and electrical conduction in syndiotactic polypropylene. *Journal of Physics. D, Applied Physics*, *33*(4), 464–471. doi:10.1088/0022-3727/33/4/321

Wang, F., Qiu, Y., Pfeiffer, W., & Kuffel, E. (2004). Insulator surface charge accumulation under impulse voltage. *IEEE Transactions on Dielectrics and Electrical Insulation*, *11*(5), 847–854. doi:10.1109/TDEI.2004.1349790

Zhang, B., Qi, Z., & Zhang, G. (2017). Charge accumulation patterns on spacer surface in hvdc gas-insulated system: Dominant uniform charging and random charge speckles. *IEEE Transactions on Dielectrics and Electrical Insulation*, *24*(2), 1229–1238. doi:10.1109/TDEI.2017.006067

Chapter 2
Flashover and Surface Charge in GIL Insulator

ABSTRACT

Many works have been studies in order to improve flashover voltage in GIL insulator. Under DC, the insulator electric field is decided by the conductivity and surface charge distribution. This chapter takes cone-type insulator as research object and then finds the characteristics of flashover, surface charge accumulation, and the interface electric field regulation (IER) of epoxy (EP)-/ graphene (GR)-coated insulator. Theoretical analysis demonstrates that the uniform surface charge of monopole is conducive by reduce peak field and flashover voltage. Among them, that of 0.1% EP/GR possesses the highest flashover voltage. With the SiC content and coating thickness enhancement of IER insulator, the electric field regulation of EP/SiC-coated insulator becomes notable, due to energy loss and increasing leakage current. The results show that insulator coated by EP/SiC can reach higher flashover voltage than uncoated insulator and enhanced SiC content contributes to improve the flashover voltage.

DOI: 10.4018/978-1-5225-8885-6.ch002

INTRODUCTION

GAS insulated lines (GILs) is considered as an ideal solution to respond to the challenge of high reliability, large capacity, compact size and environmentally friendly (Winter, 2012). Epoxy with alumina (EP/Al_2O_3) insulators are used to support conductors and provide electrical insulation, which becomes one of the key components of GILs. With the development of UHV technology and the reliability of transmission, GIL insulators must be produced in large quantities (Wang, 2017). Under operating conditions, the flashover of the GIL insulator occurs randomly (Li, 2017). In order to reduce the size of GIL insulation and improve its reliability, much research has been done on improving the flashover voltage of insulators. A new type of electrical insulating material technology can bring higher performance and reliability to power equipment and system. For example, SF6 gas insulated power equipment, such as GIS, has greatly reduced the scale and floor area of power equipment and substations. However, due to the high sensitivity of SF6 gas to the electric field, the insulation design and configuration of gas insulated power equipment become complex, and the cost of control, relaxation and optimization of the electric field is also increasing. Therefore, a new and innovative technology of electrical insulation material is one of the important tasks of future power engineering.

The main reason of flashover is caused by non-uniform field along the surface of the insulator. Therefore, changing the introduction of shielding electrodes, the optimal designs of insulator shape and embedded electrodes are introduced to solve this problem. However, these methods have greatly increased costs. Therefore, it is necessary to propose some feasible methods to solve the flashover problem of GIL insulators.

Under the action of AC and DC voltage stress, the electric field distribution along the insulator is obviously different. For a given electrode arrangement including an insulator, the electric field under AC voltage depends on the dielectric constant of the dielectric involved, thus generating a so-called capacitance field distribution. Under DC voltage, the steady-state electric field distribution, the so-called resistance field distribution, is mainly determined by the volume and surface conductivity of insulating materials. The distribution of static resistance field along the insulator is related to the accumulation of charge carriers on the dielectric interface, which may lead to the decrease of flashover voltage, especially when the polarity of applied voltage is opposite. When the DC voltage acts on the electrical insulation system, the initial capacitance field distribution will change to the fixed resistance field distribution. The transition time from one distribution to another depends on the dielectric constant and conductivity of the medium. For insulation materials with low surface conductivity and volume conductivity, such as epoxy resin, it may be within a few weeks (Gurevich, Liehr, Amiranashvili & Purwins, 2004). Because the distribution of the steady resistance field depends on the conductivity, the electrical characteristics of the dielectric involved are particularly important in the calculation. The results show that temperature and relative humidity have great influence on the volume and surface conductivity of polymer materials.

SURROUNDING INSULATING GAS

In the AC system, the regulation of dielectric constant of GIL insulator is by a great number of studies. Recently, a large number of researches has emphasized the functionally graded material (FGM) (Kato, 2006; Hayakawa, 2016). The FGM insulators that have the characteristic of spatial permittivity (ε-distribution) can be processed through 3D printing technique, the lamination and the centrifugation method. Through flashover test under lighting impulse voltage, it can be found that the breakdown voltage (BDV) of FGM insulator is significantly higher than conventional insulator.

In DC system, this paper (Nitta, 1991; Gurevich, 2004; Lutz, 2011; Du, 2018) indicates that the electric field of GILs insulator is determined by both the conductivity and the surface charge distribution, therefore, many works about the methods of surface charge accumulation mechanism and suppression have been discussed. In papers (Li, 2018), novel spacers were prepared on account of the control of adaptive surface charge, in order to enhance the flashover voltage. In paper (Shao, 2017), the enhancement of surface charge dissipation was achieved, through that the epoxy surface modified by plasma treatment. This chapter holds that the main factors that the charge accumulation causes flashover are the surface charge polarity, the insulator profile and the electrode structure. However, the surface charge accumulation can enhance flashover voltage under some circumstances. The first section (Du, 2019) takes cone-type insulator as the research object and then explore the characteristics of charge accumulation of different content EP/GR coated insulators. What's more, the function of uniform field distribution of insulator surface havs been verified, by an electrical stimulation. Finally, the insulator surface stuck with a wire-type particle to investigate the features of surface charge accumulation to metal particle.

In recent times, the ρ-nonlinear materials is regarded as a ideal method to solve the problem of uniform electric field in DC insulation (Christen, 2010; Donzel, 2011; Li, 2017; Liang, 2018; Du, 2018). The volume resistivities of ρ-nonlinear materials is affected by electric field. If the electric field decreases to a certain value, the declined volume conductivity can further reduce the electric field strength. As the result of self-regulating (IER) function of ρ-nonlinear materials, a novel method was proposed to achieve uniform field, where the IER means a technology of surface treatment. The main work of this chapter is that the EP/SiC composite coating for IER insulators are prepared, through the surface treatment of EP/SiC (Liang, 2019), at the same time, to measure the volume resistivity of EP/SiC sheet-form samples prepared. Following, the effects of SiC content and thickness of coating to the energy loss and electric field were explored by an electrical stimulation. Finally, through measuring the DC flashover voltages and coating thicknesses of coated insulator with different EP/SiC content. The EP/SiC coated insulators have two advantages as following: 1) They can reduce unpredictable distortion of electric field generated by some causes (such as defects and metal particles); 2) They may be clearly constructed by surface coating with no complicated ε/ρ-distribution, thus the manufacturing cost can be considerably reduced.

EXPERIMENTAL SETUP AND PROCEDURE

Sample Preparation

In this paper, nano graphene (GR) with different content (0.05 / 0.1 / 0.15 wt%) was dispersed in epoxy resin (EP) matrix, and 0.05, 0.1 and 0.15% EP / GR coatings were prepared. After complete mixing, the EP / GR coating is placed in vacuum for 1 hour to remove bubbles. The conical insulator is made of EP / Al2O3 composite material, and its production process is the same as that of industrial production. Before the insulator is immersed in water, the basic surface cleaning and surface roughness pretreatment must be carried out to improve the interface adhesion between insulator and coating. After soaking, drying and post-treatment, the insulator coated with EP / GR can be obtained, as shown in Figure 1. The liquid EP / GR coating was solidified into flakes by a tablet press, and the surface potential decay (SPD) and isothermal discharge current (IDC) were measured. The detailed experimental setup of SPD and IDC testing can be found in our previous work.

Surface Charge Distribution Measurement

In this paper, Kelvin type electrostatic voltmeter probe is used to measure the surface potential distribution of insulator after charging for 20 minutes under dc-20kv voltage. The insulator was fixed on the GND electrode, and the surface of the insulator was meshed into 8 × 16 cells by radial and circumferential scanning of the probe. The distance between the probe and the insulator surface is 3mm, and the signal is uploaded to the computer through the oscilloscope (Du, Liang & Li, 2019). Through inversion calculation, the measured surface potential distribution is transformed into surface charge distribution. The surface charge distribution was measured in a temperature and humidity

Figure 1. The preparation of EP/GR coated insulator

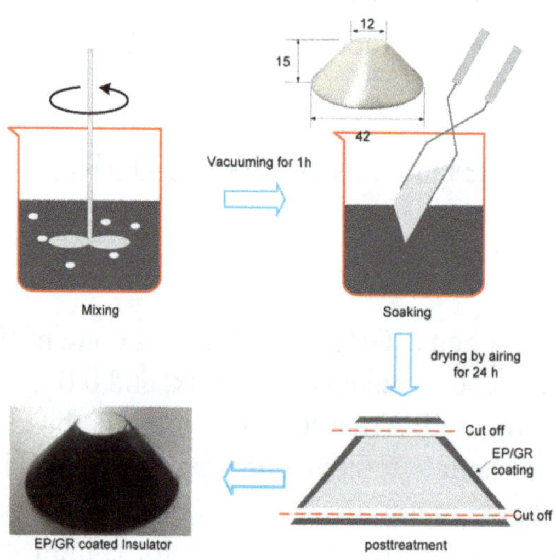

chamber. The test temperature is set at 20°C and the test humidity is set at 25%. The schematic diagram of the surface charge distribution measurement experimental device is shown in Figure 2.

Trap Energy Distribution Measurement

The trap energy distribution of EP / Al2O3 and EP / Gr Composites was measured by isothermal discharge current (IDC). First, charge at - 25kV for 30min at 50°C, then record the discharge current in short circuit at the same temperature. Based on IDC test, trap energy and trap density can be approximately calculated as follows:

$$N_t(E) = \frac{2dIt}{el^2kT} \quad (2.1)$$

$$E_t = kT\ln(\nu t) \quad (2.2)$$

Where I is the discharge current density, A/m^2; d is the thickness of the sample and l is the penetration depth of injected electrons, (m); e is the electronic charge quantity, C; k is the Boltzmann constant, J/K; T is the thermal temperature, which is set to 323K; ν is the escape frequency of trapped electrons, which is approximately equal to 10^{12} s^{-1}.

Figure 2. The experimental setup for surface charge distribution measurement

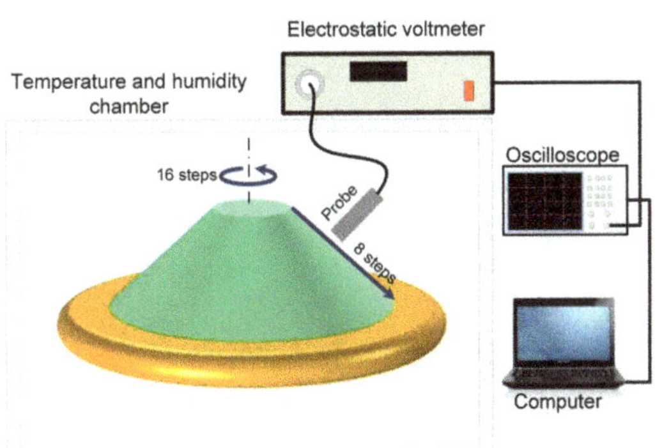

DC Flashover Test

Figure 3 is the schematic diagram of DC flashover test. The DC flashover test was carried out in a temperature and humidity chamber. The test temperature is set at 20°C and the test humidity is set at 25%. First, apply - 15 kV for 15 minutes to charge the insulator. Then the voltage is boosted at a slow action rate of 1kV / min until the flashover fault occurs. After 10 tests, the flashover voltage with 63.2% flashover probability is obtained by Weibull distribution.

Electric Simulation Model

The effect of surface charge accumulation on electric field distribution is simulated by finite element method. The simulation is performed in a rotationally symmetric model, as shown in Figure 4. The physical dimensions of insulator and electrode are given in the form of coordinates. Apply negative high voltage DC voltage to the high voltage electrode, and set the potential of GND electrode to zero. A uniformly distributed surface charge of 0 or -30 μ C / m2 is applied to the insulator surface. The electric field intensity is calculated by Poisson equation.

$$E = -\nabla V \tag{2.3}$$

$$\nabla \cdot (\varepsilon_0 \varepsilon_r E) = \rho_v \tag{2.4}$$

Figure 3. The experimental setup for dc flashover test

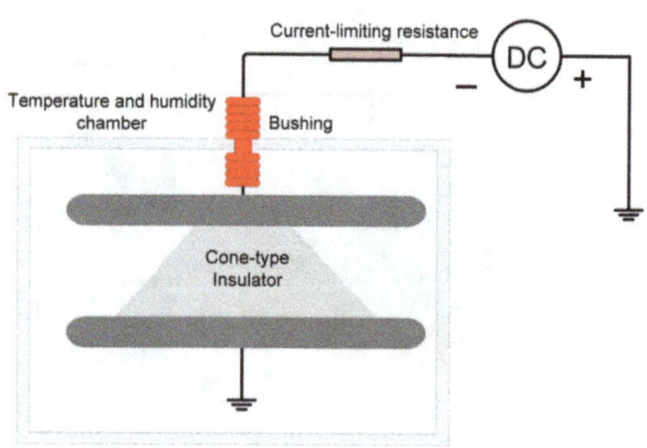

Figure 4. The simulation model for the cone-type insulator

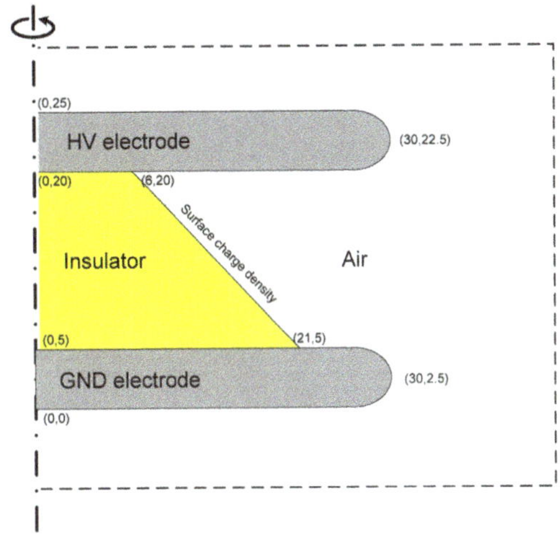

Where E is the electric field strength, V/m; V is the electric potential, V; ε_r is the relative permittivity of insulator and air; ρ_v is the space charge density, C/m^3.

FLASHOVER AND SURFACE CHARGE OF EP/GR COATED INSULATOR

Influence of Surface Charge Accumulation

Figure 5 demonstrations the Weibull distribution of the 0.1% coated and uncoated insulator flashover voltage. For 0.1% coated insulator, the flashover voltage of is -32.9 kV and increased to the cone-type insulator with 0.1% EP/GR coating.

Figure 6 illustrates the surface charge distribution of 0.1% coated and uncoated insulator below -20kV. The conclusion can be described that air ionization generates charge into surface. Affected by electric force, HV electrode attracts positive ions and GND electrode to electrons, which will migrate to the surface of insulator. The figure 6a shows that of uncoated insulator that there mainly is negative surface charge migrating to GND electrode. Influenced by electric force, the migration of negative surface

Figure 5. The Weibull distributions offlashover voltages of 0.1% coated and the uncoated insulators

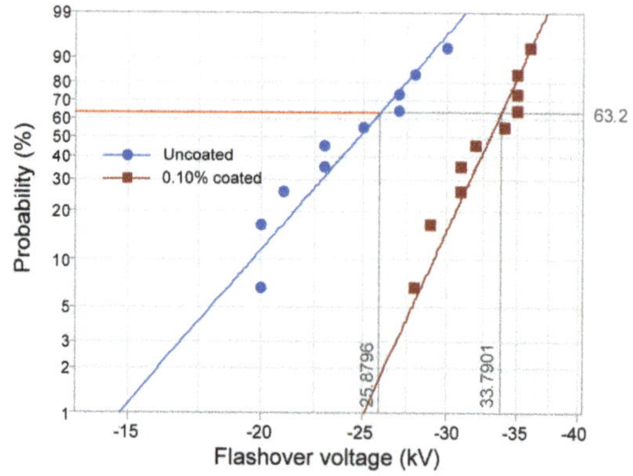

Figure 6. Surface charge distributions of the (a) uncoated and (b) 0.1% coated insulators

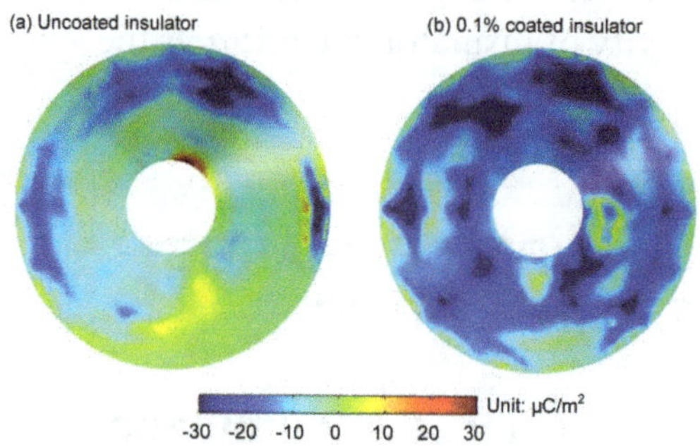

charge from the HV insulator surface to that of GND. The figure 6b displays that of the insulator coated 0.1%. It can be seen that the negative surface charge density is more uniform and keep higher. Some papers revealed that proper filler contents nanocomposites of possess SPD rates and lower carrier mobility. The EP/GR coating as a nanocomposite, own the lower carrier mobility and higher storage capacity of surface charge. Since surface charge for 0.1% coated insulator is prone to accumulate, the charge density is relatively high. The free charge is hard to migrate to the insulator surface under electric force; therefore, the distorted electric field can be weakened.

In order to explore the relationship between flashover voltage and surface charge density for the 0.1% coated insulator, how negative surface charge affects electric field distribution was analyzed by an electrical simulation. Through it, HV electrode is kept at -30 kV, GND electrode is kept at 0kV and 0 /-30 μC/m^2 is set to the surface charge of insulation. For different densities of surface charge, the Figure 7 shows the electric field distributions of cone-type insulators. The result shows that the max E-field appears at the triple junction which includes air, insulator and high voltage electrode. Hence, mass flashover faults are prone to occur. If the surface charge of insulator reach -30μC/m2, both the electric fields at the region of air and triple junction will greatly decrease, owing to the current from HV electrode which is offset by negative surface charge.

Figure 7. For different surface charge densities electric field distributions of cone-type insulators. (a) 0 μC/m2, (b) -30 μC/m2

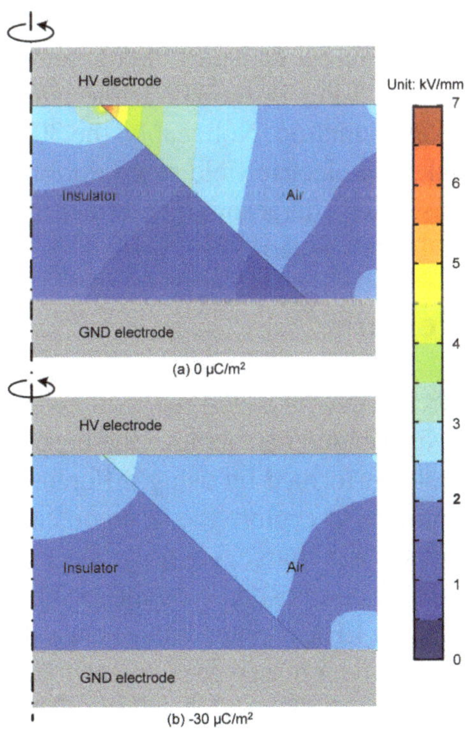

Figure 8. Electric field distributions along the creepage distance of cone-type insulator with different surface charge densities

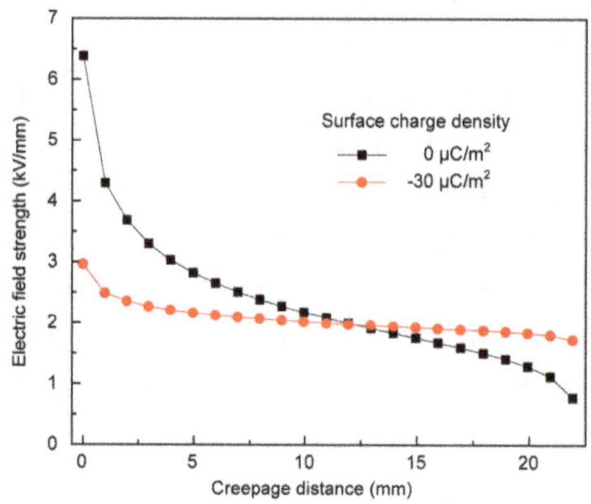

The Figure 8 shows that the electric field of cone-type insulator along the creepage surface under different densities of charge, where the *d* represents creepage, where $d = 0$ mm lies in the HV triple junction and $d = 22$ mm locates at the GND triple junction. Noticeably, the 30 μC/m^2 surface charge of insulation can make the electric field more uniform. The results shows that the single polar surface charge own the effect of field uniformity and then enhance the flashover voltage.

Effect of Filler Content

In Preceding text, the result that for cone-type insulator, the homopolar surface charge that is evenly distributed can achieve the purpose of field grading is verified. The SPD tests performed by using different filler contents on the sheet-form EP/GR samples to explore the best content of nano-graphene to enhance the density of surface charge, possibly. The normalized SPD curves of EP/GR and EP/Al2O3 composites uses the point-and-figure, shown in Figure 9. The SPD rate for EP/GR composites is less than that of EP/Al2O3 composite and reduces with increased component content, when the content of nano-graphene is below 0.1%. However, the SPD process for EP/GR composite grows quicker than that of EP/ Al2O3 composite, when the content of nano-graphene achieves 0.15%. The process of charge dissipation can be restrained by certain content of nano-graphene, the following paragraph will

Figure 9. Normalized SPD curves of EP/Al2O3 and EP/GR composites

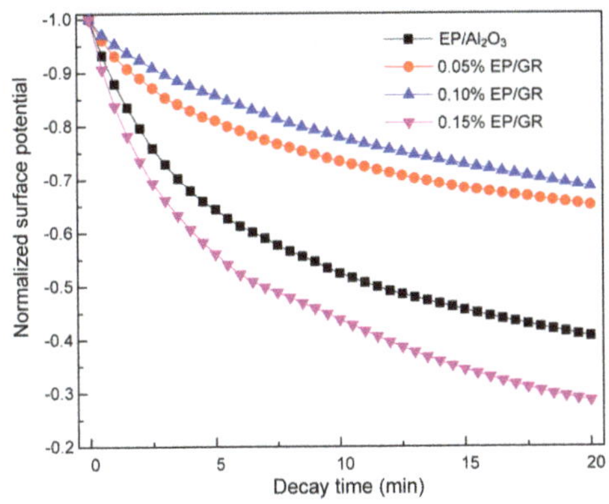

discuss. Among the other EP/Al2O3 composites and EP/GR composites, the SPD process of 0.1% EP/GR composite develops the slowest. The speculation that the highest flashover voltage and homopolar surface charge density on 0.1% coated insulator can be inferred.

The trap energy distribution of EP/GP and EP/Al2O3 composites displayed line chart of different colors obtained by IDC tests is shown in Figure 10. Owing to the deep trap existing in the interface between epoxy and nano-graphene, the charge transmission and escape can be resisted and reach a very good effect. The content of nano-graphene increases from 0 to 0.1%,

Figure 10. Trap energy distribution of EP/Al2O3 and EP/GR composites

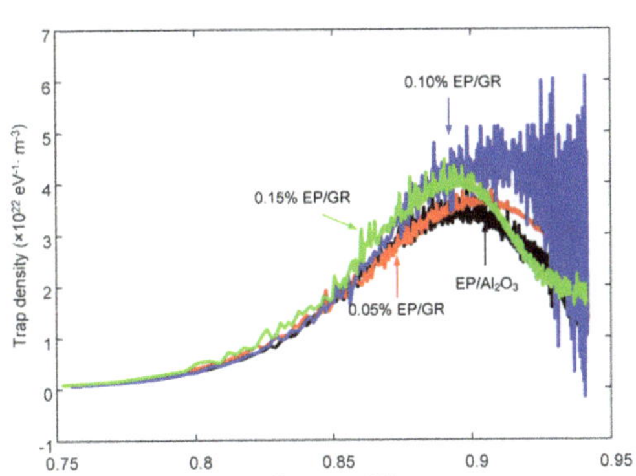

which enhances its densities of trap deep and energy, so the speed of SPD is decreased. However, the enhanced content of nano-graphene from 0.1% to 0.15% can form conductive path. The charge transport under coating is the main reason for the enhanced SPD rate. The deeper trap is easily formed for 0.1% EP / GR composite, so it has excellent charge storage capacity on the surface of insulator to other material. Table 1 lists change: carrier mobility of EP/A12O3 and EP/GR composites, through the test method mentioned in this chapter. There is the lowest carrier mobility for 0.1% EP/GR composite and the surface charge distribution is relatively uniform, because of the difficulty of surface charge transfer.

Figure 11 shows the Weibull distribution of flashover voltage for uncoated and EP/GR coated insulators. When the content of nano graphene is less than 0.1%, as the content of filler increase, the flashover voltage of uncoated insulators significantly lower than that of EP/GR coated insulators. However, the 0.15% content of nano graphene was reached, the flashover voltage of EP/GR coated insulator is lower that that of uncoated.

Table 1. Carrier mobility of EP/Al_2O_3 and EP/GR composites

GR content	EP/Al_2O_3	0.05%	0.1%	0.15%
Carrier mobility $m^2/(V{\cdot}s)$	3.8×10^{-14}	1.8×10^{-14}	1.4×10^{-14}	4.6×10^{-14}

Figure 11. Flashover voltages of the uncoated and EP/GR coated insulators

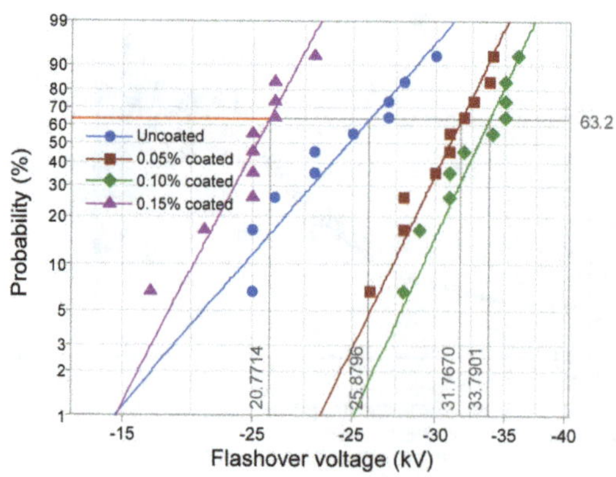

Effect of Metal Particle

During manufacturing, transportation and operation, the different impurities will be attached to the surface of insulator. Owing to the jump of metal particles between the conductor and chamber, the reliability of insulation will be reduced. The faults of flashover are prone to trigger, when the metal particles stick to the insulator surface. In this chapter, the surface of 0.1% coated and uncoated insulator are labeled by two wire-type metal particles, respectively, as shown in Figure 12. The flashover voltages of cone-type insulators effected by metal particles in air is explored, through the DC flashover tests.

The Weibull distributions of flashover voltages for the 0.1% coated and uncoated insulators under the influence of wire-type particle are shown in Figure 13. From observations, the particle tips are prone to trigger the flashover faults. The flashover voltages of uncoated and 0.1% coated insulator are reduced due to the effect of spike of wire-type metal particle. On one hand, the insulation distance is reduced by wire-type metal partial. On other hand, the tips of wire-type are prone to cause partial discharge and then form the fault of flashover. Compared to the uncoated insulator, the 0.1% coated insulator is easy to flashover, when the surface of insulator is stuck with wire-type metal. The flashover voltage is only ~-18.6 kV, for 0.1% coated insulator, but the flashover voltage for the uncoated insulator can reach ~-20.5 kV. The detailed explanation will explain in following part.

The PD easily happen at particles tips due to the small curvature radius. The serious distortion of electric field distribution is activated by the surface charge produced by PD. Figure 14 illustrates that Surface charge distribution when surface is stuck with wire-type particle, the (a) uncoated and (b) 0.1% coated

Figure 12. Cone-type insulators stuck wire-type particle on surface

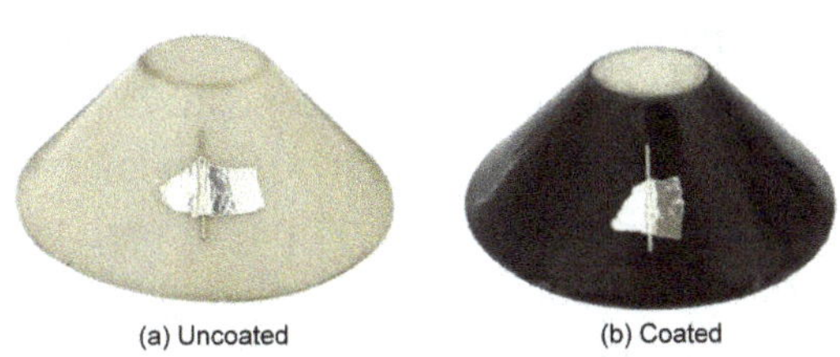

Figure 13. Weibull distribution of flashover voltages of the 0.1% coated and uncoated insulators with a wire-type particle on surface

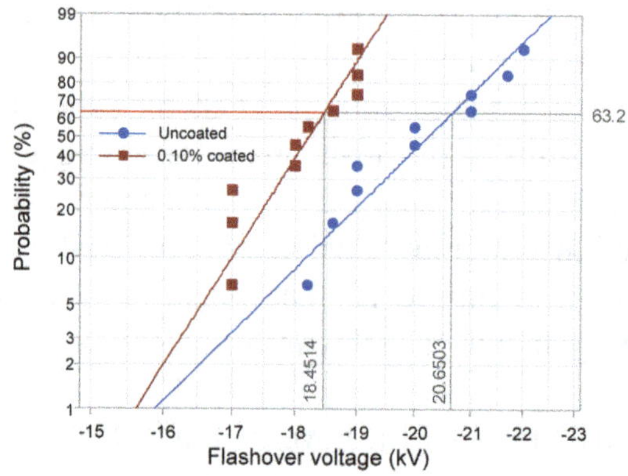

Figure 14. Surface charge distribution when surface is stuck with wire-type particle, the (a) uncoated and (b) 0.1% coated insulators

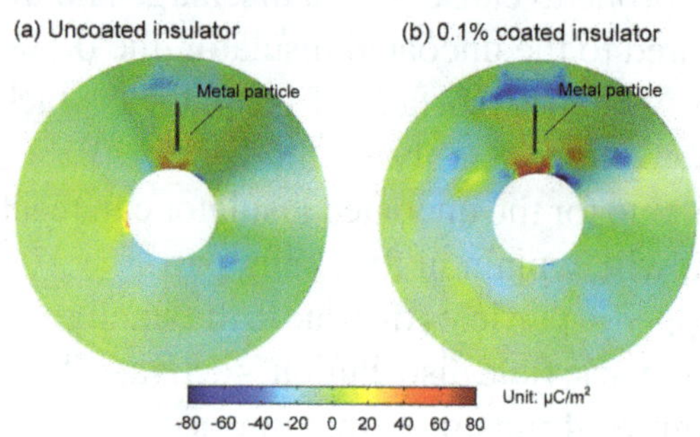

insulators under -15 kV. The positive and negative charge can be observed surrounding the tips of the HV and GND electrodes. Near the HV electrode, the tips of particle are prone to generate the positive charge spot, because of the charging of negative voltage. Similarly, the negative charge spot also will generate at the tips of G electrode. Compared with uncoated insulator, the bipolar charge spots density of the 0.1% coated insulator becomes higher, therefore the distortion of electric field should be more serious, caused by monopolar charge spot. The uncoated insulator's flashover voltage becomes higher that of the 0.1% coated.

INTERFACIAL E-FIELD SELF-REGULATING (IER) INSULATOR

Model of IER Insulator

Figure 15 illustrates the equivalent circuit model for the conventional insulator (Figure 15a), spatially nonlinear insulator (Figure 15b) and EP/SiC coated IER insulator (Figure 15c); There are many opinions provided through steady-state electric field distributions. The volume conductivity of normal insulator is regarded as a constant, owing to the EP/Al_2O_3 composites. The spatially nonlinear function can be realized. The EP/SiC composites can manufacture the spatially nonlinear insulator. It concludes two parts that the body is made of EP/Al2O3 composites and the skin layer is made of EP/SiC composites. Figure 15a, by differentiation, a cone-type insulator can be separated into n→∞ elements of cylinder, where the r_i is the radius and dL is the height, the R_i is the resistance and the ΣR_i can be regarded as series resistance of insulator. The resistance of element can be computed by equation (3-1):

$$R_i = \rho \frac{dL}{\pi r_i^2} \tag{4-1}$$

where in steady state, the electric potential drop dV_i, ρ is the resistivity of the insulator, Ω·m. on element i can be obtained according to the Ohm's law as:

$$dV_i = I_c R_i \tag{4-2}$$

where Ic is the steady-state conductive current, A. Based on the above analysis, the E-field strength on element i can be described as:

$$E_i^a = \frac{dV_i}{dL} = \frac{I_c}{\pi} \frac{\rho}{r_i^2} \tag{4-3}$$

According to Equation (4-3), the E-field distribution along the surface of cone-type insulator is tremendously inhomogeneous. The maximum E-field strength lies at HV electrode and decreases from HV to GND electrode.

The figure 15b shows the spatially nonlinear insulator, it can be seen that the volume resistivity is decided by electric field. The electric field can be reduced by the rapidly decrease volume resistivity, when electric field reaches

Figure 15. Equivalent-circuit models for (a) the conventional insulator, (b) the spatially nonlinear insulator and (c) the IER insulator

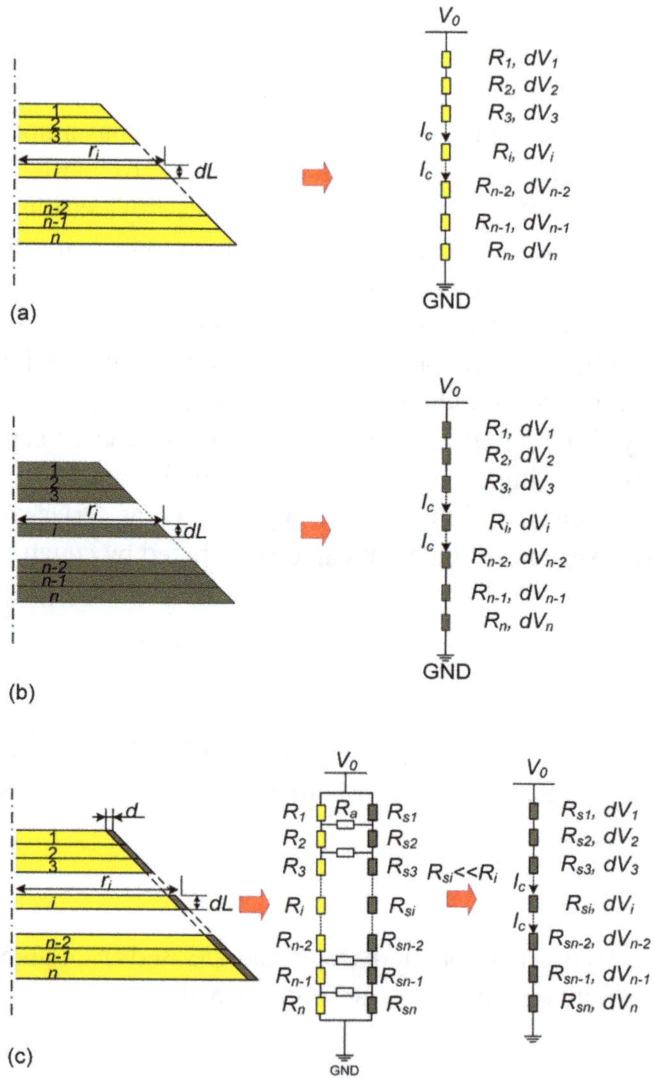

a definite value. The figure 15 the electric field of element i of the spatially nonlinear insulator can be represented by equation (4-4):

$$E_i^b = \frac{I_c}{\pi}\frac{\rho(E)}{r_i^2} \tag{4-4}$$

where $\rho(E)$ is the E-dependent volume resistivity of the spatially nonlinear insulator, Ω·m. Compared the electric field distribution of conventional insulator, that along the spatially nonlinear insulator will be slightly uniform, owing to the function of E-field regulation of $\rho(E)$.

The figure 15c shows the insulator coated by EP/SiC. Both the insulator and skin layer can be divided into n ($n \to \infty$) elements. Each element i of the skin layer can be considered as a column ring with a radius of r_i, a thickness of d and a height of dL. The resistance value of element i of the skin layer can be approximately calculated as:

$$R_{si} = \rho_s(E)\frac{dL}{2\pi r_i d} \tag{4-5}$$

where $\rho_s(E)$ stands the E-dependent volume resistivity of the skin layer, Ω·m. In the equivalent-circuit model for the EP/SiC coated insulator, the body resistor series ΣR_i and the skin layer resistor series ΣR_{si} are combined in parallel through a contact resistance R_c. When $R_{si} << R_i$, the equivalent circuit can be simplified to the skin layer resistor series ΣR_{si} only. The E-field strength on element i of the EP/SiC coated insulator can be illustrated as:

$$E_i^c = \frac{I_c}{2\pi d}\frac{\rho_s(E)}{r_i} \tag{4-6}$$

The E-field distribution along the surface of EP/SiC coated insulator should be more uniform than that along the spatially nonlinear insulator surface because E_i^c is proportional to $\rho_s(E)/r_i$ and E_i^b is proportional to $\rho_s(E)/r_i^2$, as shown in Equations (4-4) and (6).

Electric Field Distribution

The tangential electric fields of space nonlinear insulator and IER insulator coated by EP/SiC are computed, in order to discuss the effect of the coating thickness, as illustrated in Figure 16. From the simulation, the SiC contents of spatially nonlinear insulator and EP/SiC coated insulators are set to 20%, where the creepage distance d = 0 mm represents the position of HV electrode

Figure 16. E-field distribution along the conventional insulator, IER insulators and spatially nonlinear insulator

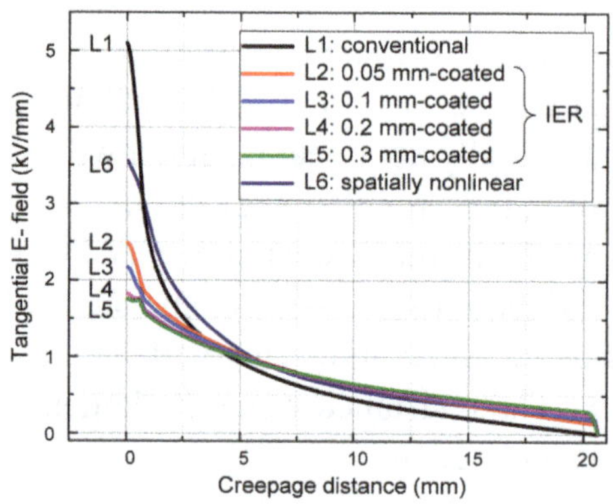

and d = 21 mm lies in the GND electrode. The electric field is prone to concentrate at the region of HV triple junction, so the flashover fault will develop from the HV triple junction to the GND triple junction. Compared with the maximum field ~5kV/mm of conventional insulator, the maximum field of spatially nonlinear insulator can decrease to ~3 kV/mm, owing to the regulation function of electric field for EP/SiC composite. Evidently, the enhanced coating thickness of EP/SiC insulator will uniform the electric field distribution. However, more than 2mm thickness can generate the tiny influence to the electric field distribution.

There is the "saturation effect" for the electric field regulation, which can be explained: the resistance R_{si} of element i will decrease with the enhancement of thickness d.

Assuming that (RSI < RI) holds under certain conditions, the tangential electric field strength is more likely to obey equation (6), for the thicker EP / SiC coating. Assuming that (RSI < < RI) also holds, when the coating thickness reaches a certain value, the continuous growth of the coating thickness has little effect on the electric field distribution of EP / SiC composite coating insulator.

During the operation of insulator, the energy loss is mainly caused by leakage current. Figure 17 shows the loss power density distribution pin for conventional insulator, EP / SiC coated ier insulator and spatial nonlinear insulator. Figure 17a shows that for traditional insulators, the energy loss

Figure 17. Loss power density distributions of (a) the conventional insulator, (b) the IER insulator and (c) the spatially nonlinear insulator

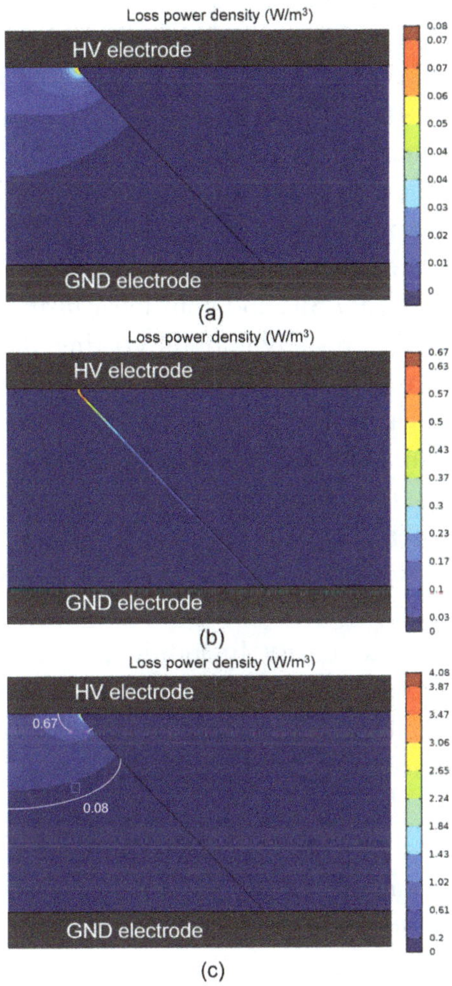

in the area at the three nodes of high voltage is more concentrated, and the electric field is also more concentrated. Among them, the maximum power loss density is as low as 0.07 w / m3, so conventional insulator has good insulation performance for power loss. Figure 17b shows the EP / SiC coating insulator. Due to the small resistivity of the coating, the leakage current is large, so the energy loss in the coating is large, which can reach 0.67 w / m3. For the space nonlinear insulator shown in Figure 17, because the insulator is made of EP

/ SiC composite insulation material, the low surface resistivity and volume resistivity lead to large leakage current, so its maximum power loss density even reaches 4.04w/m3, about 5 times of the traditional insulator. In order to compare the maximum power loss density of conventional insulator (0.08 w / m3) and EP / SiC coated insulator (0.67 w / m3), its isoline is marked in Figure 17c. The loss power density of space nonlinear insulator is higher than that of conventional insulator and EP / SiC coating insulator in region 1 and 2, respectively. The energy loss of space nonlinear insulator is too large to realize in practical application.

Figure 18 shows the leakage current and loss power of traditional insulator, er insulator coated with EP / SiC and non-linear insulator based on integral calculation. Due to the low resistivity of the coating, the leakage current and power loss density of the space nonlinear insulation are higher than that of the treated insulator. In addition, the surface electric field regulation function of space nonlinear insulator is not as effective as that of the insulator coated with EP / SiC. The results show that EP / SiC coating insulator is more suitable for power system, compared with other insulators. According to the characteristics of resistivity, it is easy to know that when the thickness of the coating increases, the leakage current will inevitably increase, resulting in the increase of power loss density. In addition, the electric field regulation function shows "saturation effect", when the coating thickness is more than 0.2mm. Therefore, the best coating thickness is 0.2mm when considering the effect of capacity loss and electric field regulation.

Figure 18. The leakage current and loss power of the conventional insulator, IER insulators and spatially nonlinear insulator

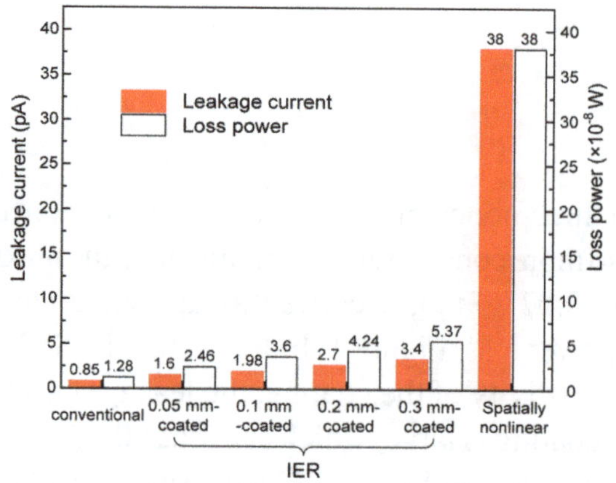

Figure 19 shows the tangential electric field for conventional insulators with different SiC content and ier insulators coated with EP / SiC. In the simulation, for the EP / SiC coating insulator, the coating thickness is determined as 0.2mm, and the SiC content is controlled between 15% and 25%. However, with the increase of SiC content, the nonlinear threshold of EP / SiC coating will decrease. The maximum electric field of EP / SiC coating insulator is near the lower value of nonlinear threshold. Therefore, the higher the SiC content, the more uniform the electric field. However, it's not that the thickness has been increasing about well. Because the high content of SiC will reduce the resistivity and increase the leakage current, resulting in higher loss power, as shown in Figure 20. At the same time of improving the electric field, the leakage current caused by the coating is an adverse effect. Therefore, in order to keep the loss power density within the allowable range, the insulator with appropriate SiC content can reduce the electric field distortion.

Due to the large DC insulation resistance in DC system, its steady-state electric field distribution usually takes days, weeks or even months to achieve. In the previous paper, we have calculated the steady-state electric field.

In the actual operation process, the electric field is often transient, so we establish a transient simulation model. In the simulation, the coating thickness is set to 0.2mm, and the SiC content is set to vary from 15% to 25%. In 20 minutes (the first stage), the applied voltage is increased from 0 kV at a uniform speed of 0.5 kV / min, and then kept constant in 2018 minutes (the second stage), as shown in figure 21a. According to the previous analysis

Figure 19. E-field distribution along the IER insulators with different SiC contents

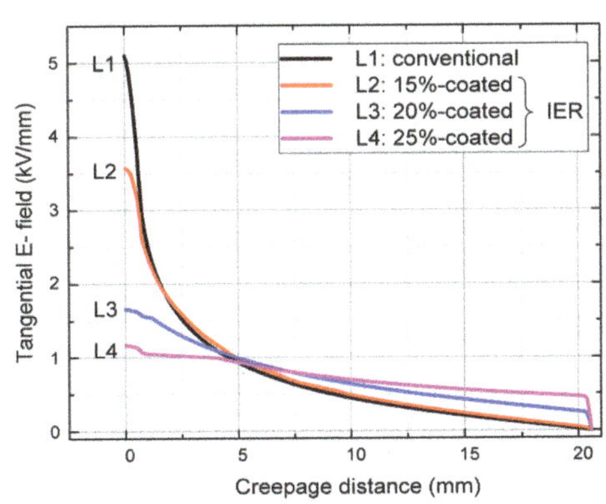

Figure 20. Leakage current and loss power of the IER insulators with different filler contents

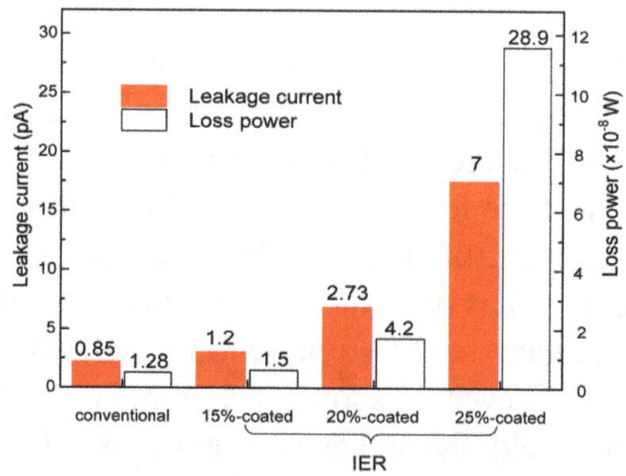

Figure 21. Transient process of (a) the voltage rising and (b) the maximum E-field strength variation

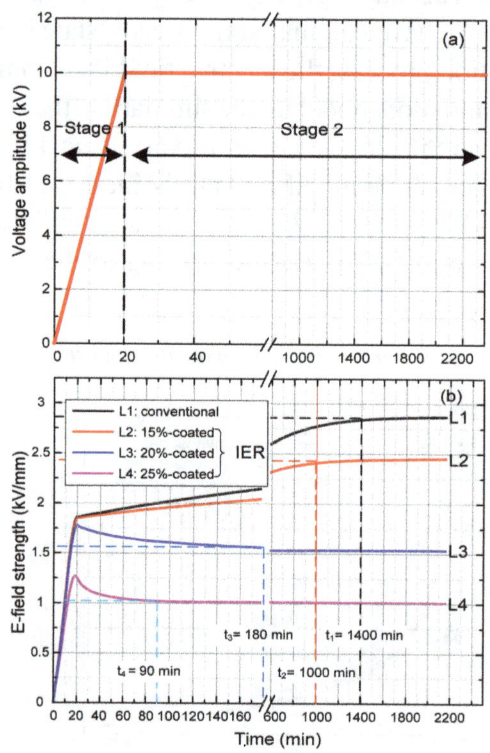

results, the maximum electric field of any insulator appears at the junction of HV and HV. Figure 21b shows the maximum field strength change by comparing the conventional insulator with the IE insulator covered by EP / SiC. In the first stage, the maximum field strength of 15% coating insulator and conventional insulator increases continuously. Even in the second stage, when there is a constant voltage, the electric field will tend to a stable value over time. According to the simulation results, the steady-state establishment time of conventional insulator and 15% coating insulator is 1400 and 1000 minutes. The maximum field strength of 20% and 25% coating insulator increases first and then decreases with time, and then reaches a stable value. The steady-state set-up time of 20% and 25% claddings is 180 and 90 minutes respectively, which is greatly reduced compared with the former. According to the theory of electromagnetic field, this is because the insulation resistance is greatly reduced. The lower the volume resistivity is, the faster the charge transfer process is. Because the volume resistances of 20% and 25% coated insulators are lower, the fast distribution of space charge results in faster steady-state build-up time. Therefore, the steady-state build-up time of EP / SiC coating insulator will be greatly reduced, with the increase of SiC content.

CONCLUSION

Under DC voltage, the electric field distribution depends on not only the conductivity of the dielectrics but also the charge distribution on the insulator surface. This chapter introduced two methods to regulate the electric field distribution along the insulator surface. The EP/GR coated insulators with a proper GR content can increase the homopolar surface charge accumulation, which reduces the electric field strength on the insulator surface. By experiment, the 0.1%-coated insulator can increase the flashover voltage by 30% compared with that of the conventional insulator. The nonlinear conductivity material has the function of regulating the DC electric field distribution. By coating EP/SiC composites with different SiC contents, the IER insulators has can intelligently reduce the electric field distortions on insulator surface. By increasing the coating layer thickness and SiC content, the IER insulator gains better electric field regulating effects and higher flashover voltages but larger leakage currents. In practical application, a proper coating layer content and SiC content is required to gain the best electric field regulating effect but the lowest leakage current.

REFERENCES

Christen, T., Donzel, L., & Greuter, F. (2010). Nonlinear resistive electric field grading part 1: Theory and simulation. *IEEE Electrical Insulation Magazine*, *26*(6), 47–59. doi:10.1109/MEI.2010.5599979

Donzel, L., Greuter, F., & Christen, T. (2011). Nonlinear resistive electric field grading Part 2: Materials and applications. *IEEE Electrical Insulation Magazine*, *27*(2), 18–29. doi:10.1109/MEI.2011.5739419

Du, B. X., Liang, H. C., & Li, J. (2019). Surface coating affecting charge distribution and flashover voltage of cone-type insulator under DC stress. *IEEE Transactions on Dielectrics and Electrical Insulation*, *26*(3), 706–713. doi:10.1109/TDEI.2019.8726015

Du, B. X., Liang, H. C., & Li, J. (2019). Interfacial E-field self-regulating insulator considered for DC GIL application. *IEEE Transactions on Dielectrics and Electrical Insulation*, *26*(3), 801–809. doi:10.1109/TDEI.2019.8726027

Du, B. X., Liang, H. C., Li, J., & Zhang, C. (2017). Temperature dependent surface potential decay and flashover characteristics of epoxy/SiC composites. *IEEE Transactions on Dielectrics and Electrical Insulation*, *25*(2), 631–638. doi:10.1109/TDEI.2017.006872

Gurevich, E. L., Liehr, A. W., Amiranashvili, S., & Purwins, H. G. (2004). Role of surface charges in dc gas-discharge systems with high-ohmic electrodes. *Physical Review. E*, *69*(3), 036211. doi:10.1103/PhysRevE.69.036211 PMID:15089396

Hayakawa, N., Ishiguro, J., Kojima, H., Kato, K., & Okubo, H. (2016). Fabrication and simulation of permittivity graded materials for electric field grading of gas insulated power apparatus. *IEEE Transactions on Dielectrics and Electrical Insulation*, *23*(1), 547–554. doi:10.1109/TDEI.2015.005237

Kato, K., Kurimoto, M., Shumiya, H., Adachi, H., Sakuma, S., & Okubo, H. (2006). Application of functionally graded material for solid insulator in gaseous insulation system. *IEEE Transactions on Dielectrics and Electrical Insulation*, *13*(2), 362–372. doi:10.1109/TDEI.2006.1624281

Li, C., Hu, J., Lin, C., & He, J. (2017). The potentially neglected culprit of DC surface flashover: Electron migration under temperature gradients OPEN. *Scientific Reports*, *7*(1), 3271. doi:10.103841598-017-03657-1 PMID:28607418

Li, C., Hu, J., Lin, C., Zhang, B., Zhang, G., & He, J. (2017). Surface charge migration and dc surface flashover of surface-modified epoxy-based insulators. *Journal of Physics. D, Applied Physics*, *50*(6), 065301. doi:10.1088/1361-6463/aa5207

Li, C., Lin, C., Hu, J., Liu, W., Li, Q., Zhang, B., ... He, J. (2018). Novel HVDC spacers by adaptively controlling surface charges – part i: Charge transport and control strategy. *IEEE Transactions on Dielectrics and Electrical Insulation*, *25*(4), 1238–1247. doi:10.1109/TDEI.2018.007054

Li, J., Du, B. X., Kong, X. X., & Li, Z. L. (2017). Nonlinear conductivity and interface charge behaviors between LDPE and EPDM/SiC composite for HVDC cable accessory. *IEEE Transactions on Dielectrics and Electrical Insulation*, *24*(3), 1566–1573. doi:10.1109/TDEI.2017.006198

Li, Z. L., Du, B. X., Yang, Z. R., & Li, J. (2017). Effects of crystal morphology on space charge transportation and dissipation of SiC/silicone rubber composites. *IEEE Transactions on Dielectrics and Electrical Insulation*, *24*(4), 2616–2625. doi:10.1109/TDEI.2017.006388

Liang, H. C., Du, B. X., Li, J., Li, Z. L., & Li, A. (2018). Effects of Nonlinear Conductivity on Charge Trapping and De-trapping Behaviors in Epoxy/SiC Composites under DC Stress. *IET Science, Measurement & Technology*, *12*(1), 83–89.

Lutz, B., & Kindersberger, J. (2011). Surface charge accumulation on cylindrical polymeric model insulators in air: Simulation and measurement. *IEEE Transactions on Dielectrics and Electrical Insulation*, *18*(6), 2040–2048. doi:10.1109/TDEI.2011.6118642

Nitta, T., & Nakanishi, K. (1991). Charge accumulation on insulating spacers for HVDC GIS. *IEEE Transactions on Electrical Insulation*, *26*(3), 418–427. doi:10.1109/14.85113

Shao, T., Liu, F., Hai, B., Ma, Y., Wang, R., & Ren, C. (2017). Surface modification of epoxy using an atmospheric pressure dielectric barrier discharge to accelerate surface charge dissipation. *IEEE Transactions on Dielectrics and Electrical Insulation*, *24*(3), 1557–1565. doi:10.1109/TDEI.2017.006321

Wang, H., Guo, Z., Feng, H., Liu, L., Tian, H., Wang, C., & Peng, Z. (2017). Simulation and experimental study on mechanical performance of UHV GIS spacer in hydrostatic test. *IEEE Transactions on Dielectrics and Electrical Insulation*, *24*(5), 3122–3131. doi:10.1109/TDEI.2017.006371

Winter, A., & Kindersberger, J. (2012). Stationary resistive field distribution along epoxy resin insulators in air under DC voltage. *IEEE Transactions on Dielectrics and Electrical Insulation*, *19*(5), 1732–1739. doi:10.1109/TDEI.2012.6311522

Chapter 3
Treeing Characteristics in HTV Silicone Rubber

ABSTRACT

Silicone rubber (SiR) is widely used as the main insulating material in cable accessories and faces high temperature challenges during operation. Pulse overvoltages can threaten working state of the insulation. In order to understand electrical tree degradation process at high temperature under repeated pulse voltages, stress testing was performed in this experiment. The ambient temperature (Tamb) was set to 30, 60, 90, 120, and 150°C, and the frequencies of the pulses were set to 5, 20, 100, 200, and 1000 Hz, respectively. During operation, it is also threatened by mechanical stress, which is caused by the spring-clamping device and the expanded bundling force of the rubber stress cone, which will affect the electrical properties of the silicone rubber in the cable accessory. In this chapter, the growth characteristics of trees under tensile stress and compressive stress were studied by using the needle plane electrode system.

DOI: 10.4018/978-1-5225-8885-6.ch003

INTRODUCTION

During abnormal operation, high temperature caused by overloaded load current of the conductor in cable can seriously threaten the stable operation of the insulation. The temperature gradually decreases. For inner conductor, the rated load temperature is about 90°C, and the operating temperature is usually designed to be 50 to 60°C, during the transient failure of the extruded XLPE cable system, the temperature during operation may rise to 150°C. In (Chen, 2015), the electric tree characteristics of XLPE cables at different temperatures were studied in detail. However, research on the insulation properties of cable accessories is currently less relevant at different high temperatures. Electrical trees are the phenomenon of the occurrence and propagation of electrically induced tubes, which are the main electro-degradation phenomena in dielectric insulation materials (Dissado, 2002). However, the degradation mechanism of cable accessories at different high temperatures has not been fully studied. Connector failures accounted for about 21% of the cable system. According to the statistical results of the national grid for 110 kV and above cable systems, more than 70% of cable faults originate from faults in cable accessory equipments. Therefore, based on the practical view, it is of extraordinary significance to explore the phenomenon of electric tree initiation and the propagation process in silicone rubber.

SiR is employed as the main insulating materials for the cable accessory and is main insulating materials for cable joints as well as the cable terminals. In order to ensure certain electrical strength of the interface, the cable attachment is coated to the main insulation layer under the certain amount of interference force which generated by the spring clamp device as well as the silicone rubber stress cone. However, mechanical stress can easily cause deformation of the material during installation of the cable attachment and during changes in the operating environment. Therefore, it is inevitably subjected to the varying mechanical stresses for the silicone rubber.

Mechanical stress and strain are relevant to the electrical property of polymer material, which affect the electrical breakdown strength of the materials. It shows that the breakdown field decreases when the strain is up to 30%, and increases when the strain is up to 160%. Electrical tree is the main cause of electrical dilapidation and breakdown of insulating materials. According to state grid statistics, more than 70 percent of cable failures were caused by failures of cable accessories in cable systems 110 kV and above. Electrical trees usually originate from electric field focus points, such as impurity, air gaps or mechanical defect (Wu, 2005; Dissado, 2002; Shimizu, 1998; Eichhorn, 1977). The propagation characteristics of electrical trees are influenced by a lot of factors, such as the voltage waveform, electromechanical stress, operating temperature, chemical decomposition and physical distortion (Ahmad, 1989). Mechanical stress would also affect growth characteristics of trees in electrical insulation (Varlow, 2002). Al-ghamdi et al. (2004) reported the growth process of electrical tree in epoxy resin under mechanical pre-stress, and found that the growth of trees slowed down due to compression stress, and a longer growth time of trees was observed (Ghamdi, 2004).

The effect of different mechanical stresses on the growth characteristic of HTV silicone rubber trees were studied. In the paper, the growth process of electric tree was observed using a digital microscope. The morphology and growth characteristics of electrical trees under different compressive and tensile stresses were studied. Accumulated damage and Fractal dimension were used to analyze the properties of electrical tree. In order to figure out the influence of mechanical stress on the initiation characteristics of electrical trees, the initiation probability and initiation time of electrical trees are also studied. In addition, the formation and development mechanism of electric branches under tensile and compressive stress is also discussed.

TREEING PROPERTY IN SIR UNDER VARIOUS TEMPERATURE AND MECHANICAL STRESS

Temperature-Dependent Electrical Tree in Silicone Rubber Under Repetitive Pulse Voltage

In the experiment, repeated pulse voltage was applied to samples at different temperatures, and the typical electrical tree morphology under diverse conditions has been shown in Figure 1. Tamb in Figure 1a and Figure 1b is 30°C and 90°C. In this figure, the dendritic tree (a1) was obtained at 9kV, the dendritic tree (a2) was observed at 12kV, and the stem tree (a3) was observed at 15kV. In Figure 1b, with the increase of pulse amplitude, the structure of the electric tree will also show great differences. The results show that the growth process was related to the amplitude of the pulse.

Figure 1. Typical tree morphology at different conditions

The typical tree form is shown in table 1. The pulse frequency is 20Hz, and the number of pulses (n-pulse) is 10000.As shown in the figure, pulse voltage amplitude, pulse frequency and temperature all have a great influence on the shape of electrical branches. When you apply 15kV, Tamb is less than 90°C, the branching tree is the main tree. Figure 2 shows the typical morphological distribution at different pulse frequencies. Shrubs increased from 0% to 100% as pulse frequency increased, while branches showed an opposite trend. The results show that the pulse voltage and pulse temperature not only affect the shape of the tree, but also affect the shape of the tree.

Figure 3 shows the typical shape of the tree at different pulse frequencies when Npulse = 50. It is worth noting that the total number of pulses is a certain value. In Figure 3a, when the pulse frequency increased to 200Hz, only one branch tree was observed, with a slight increase in length. One branch of discharge energy is limited when Tamb is 30°C, so it is difficult to lower the main branch around the 9 kV silicone rubber chain alone. However, when Tamb rises to 90°C, the tree channel will increase by 9 kV pulse frequency, but

Table 1. Typical electrical tree morphology at different temperature and pulse amplitude

	9 kV	12 kV	15 kV
30 °C	branch/bush	branch/bush	branch
90 °C	branch/bush	branch/bush	branch
150 °C	bush	bush	bush

Figure 2. Distribution of typical electrical tree morphology with different pulse frequencies

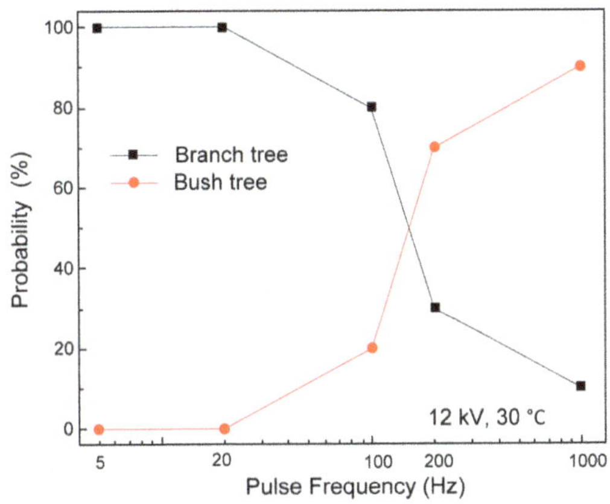

Figure 3. Typical tree morphology at different pulse frequencies for Npulse= 50

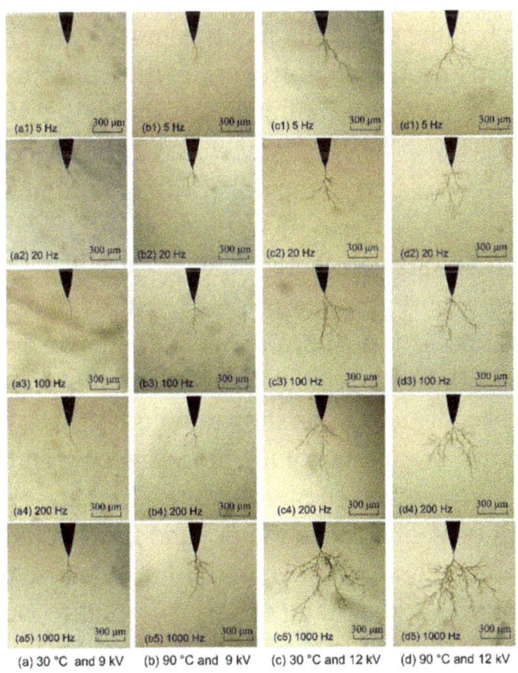

this trend is more obvious 12 kV, which shows that the characteristics of the tree are related to the pulse frequency. As shown in Figure 3c and Figure 3d, with the increase of pulse voltage frequency, the bifurcation of the electrical branch channel gradually increases, and its length also increases, the damage area also increases, and the damage degree to the material is more serious. Under various conditions, with the increase of pulse frequency, the electrical branch channels show the same change trend.

The tree length is used to describe the growth of the tree in the direction of the needle electrode. The length of the electric branch directly describes the resistance and penetration characteristics of the insulation material. When the electric branch channel develops to the ground electrode, the insulation will face the risk of breakdown. The filling factor of two-dimensional projection real electric tree mode is described with FD (Kudo, 1998; Mason, 1955).In this way, the morphology and structure of electric branches can be described more comprehensively. Figure 4 illustrates the relationship between tree length (a) and FD (b) and pulse frequency when Npulse = 10000. At 9kV, when the pulse frequency increases from 20Hz to 1000Hz, the length of the tree increases slightly, but FD is significant. The tree is associated with an electric field, and when the tree's channel tips move away from the tip of the needle, it almost stops growing. However, as the pulse voltage on the test sample is maintained, the number of channels in the tree increases, leading to an increase in FD. It can be clearly seen from the Figure that the length of the electric branch is related to the pulse voltage frequency of the fractal dimension, and with the increase of frequency, the length of the electric branch and the fractal dimension both increase rapidly.

In the test, the number of pulses was 10,000. At 30°C, a typical branching tree structure (a1) was gained and the tree was sparse tree In the test, the number of pulses was 10,000.30 °C, are typical of the branching tree structure (Figure 5 a1), for the sparse dendritic tree, as shown in Figure 5 a2, with temperature increasing, the electric branch becomes more intense, and the length less, as shown in Figure 5 a3, when the temperature higher, the electric branch near the needlepoint present a dense jungle, and the length of the narrowed. With the increase of temperature, the length of the electric branch gradually shrinks and the damage degree to the material gradually increases, which is manifested by its color gradually becoming darker. This trend is shown in Figure 5 a, b and c, and this trend is most obvious in Figure 5c.

The passage can be clearly observed. Typical shrub tree structures (Figure 5 c3) were detected at 150°C, with the tree channels closely interconnecting with each other. With an increase in pulse frequency, the structure of shrub

Figure 4. Tree length (a) and FD (b) versus pulse frequency for Npulse= 10000

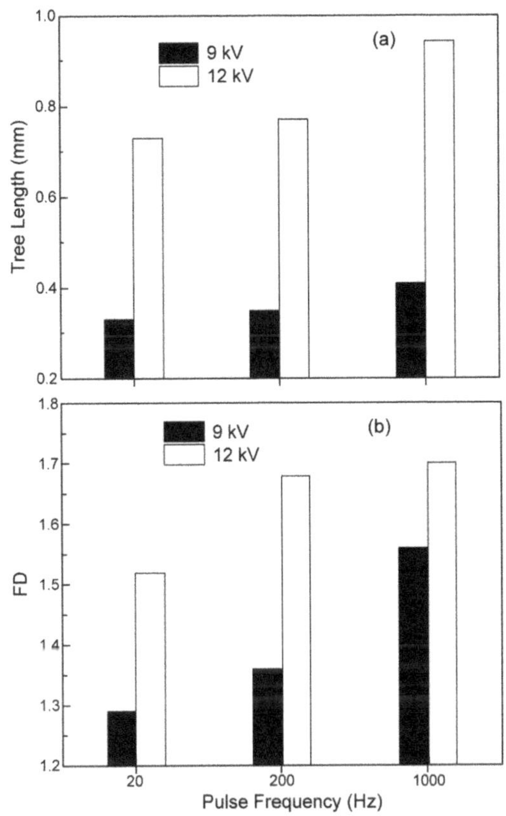

trees also changed. In (Figure 5 a3) only the "black area" of the electric tree was observed around the tip of the pin, while in (Figure 5 c3) almost the entire tree area was filled.

Figure 6 displays the relationship between length (a) and FD (b) of the tree at different temperatures and the number of pulses. The pulse amplitude was 12 kV and the pulse frequency was 200 Hz. Compared with 30°C, the tree with Npulse = 10000 is shorter at 150°C, even though it is longer at the beginning. However, as Tamb changes from 30°C to 150°C, FD continues to increase. The results are consistent with the tree witnessed in Figure 6. Figure 6 shows the relationship between the length (a) and FD (b) of the tree at different temperatures and the number of pulses. The pulse amplitude is 12kV and the pulse frequency is 200Hz.Compared to 30°C, trees with Npulse

Figure 5. Typical shapes of electrical trees at different temperatures for Npulse= 10000 at 12 kV

= 10000 were shorter at 150°C, although they started out longer. However, FD continues to increase as Tamb changes from 30°C to 150°C.The result is consistent with the tree in Figure 6. As shown in Figure 6 a, when the number of pulses is small, the length of the electric branch increases gradually with the increase of temperature, but when the number of pulses is large, the length of the electric branch decreases monotonously with the increase of temperature. Other things being equal, the length of the branch increases rapidly with the number of pulses. This is because the number of partial discharges in the branch channel increases with the number of pulses, which releases more energy and breaks more molecular chains. As shown in Figure 6b, with the increase of temperature, the fractal dimension monotonically increases. This is because with the increase of temperature, electrical branches gradually change from dendritic to clumpy, and their density gradually increases, so the fractal dimension monotonically increases.

Figure 7 displays the relation between the length (a) and FD (b) of the tree and the temperature under different pulse amplitude. The stress pulse frequency is 200 Hz, and the number of pulses is 10000. When Tamb increases from 30 to 150°C, the tree length decreases slightly, while when the pulse amplitude changes from 9kV to 15 kV, the tree length significantly increases. When

Figure 6. Tree length (a) and FD (b) versus pulse number with application 12 kV pulse voltage at 200 Hz

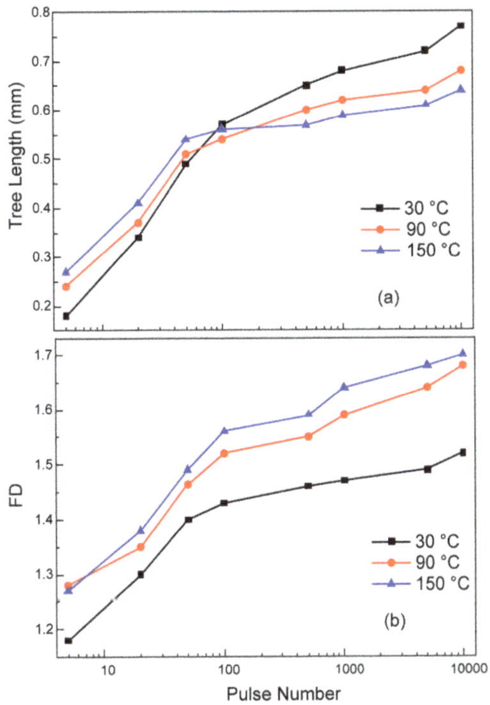

Tamb changes from 30°C to 150°C, FD continues to rise, which is relative to the probability of increasing shrub tree shape at high temperature. When sample is subjected to a stress of 15 kV, the FD is less than 12 kV belows 90°C, and is larger at 12 kV below 150°C. The results is related to the change of tree morphology at 15 kV. Figure 7 shows the relationship between the length (a) and FD (b) of the tree shape and temperature under different pulse amplitudes. The stress pulse frequency is 200Hz and the number of pulses is 10000.When Tamb increases from 30°C to 150°C, the length of the tree decreases slightly, while when the pulse amplitude increases from 9kV to 15kV, the length of the tree increases significantly. When Tamb changes from 30°C to 150°C, FD continues to increase, which is related to the probability of increasing the shape of shrubs and trees at high temperatures. When the specimen withstands the stress of 15 kV, FD is less than 12 kV and less than 90°C, and greater than 12 kV and less than 150°C. This result is related to the

Figure 7. Tree length (a) and FD (b) versus temperature at 200 Hz

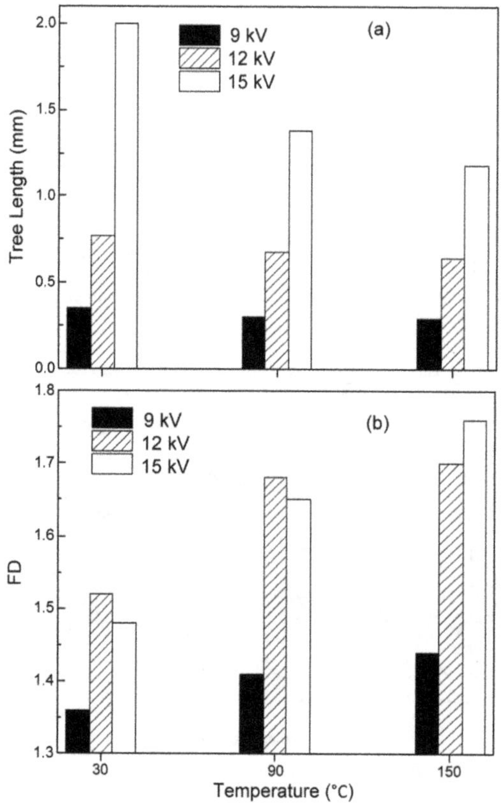

change of tree morphology at 15kV. Figure 8 shows the relationship between the pulse frequency and the probability of electrical branch initiation, in which the voltage is 9kV and the number of pulses N=50. As can be seen from the figure, the probability of initiation increases monotonically as the pulse frequency increases, which will be discussed in the following chapters.

The probability of tree breakdown was investigated with Npulse = 10000 at different temperatures, and the results were shown in Figure 10. The probability of electric tree breakdown decreases with the increase of temperature, but it is independent of pulse frequency. Notice that when Tamb goes from 30 to 150°C, the breakdown probability rises from almost 90% to 30% at 20 Hz.

In Figure 1.b, the results show that pulse amplitude play an vital role during the growth process of electrical trees. Mason's formula (Mason, 1955) can be used to calculate the electric field created by the pulse voltage at needle tip:

Figure 8. Relationship between the inception probability and the pulse frequencies with application 9 kV for Npulse= 50

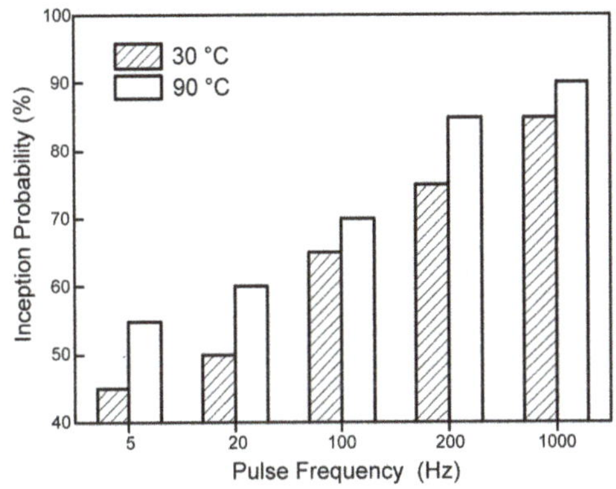

Figure 9. Relationship between the inception probability and the pulse number at 20 Hz

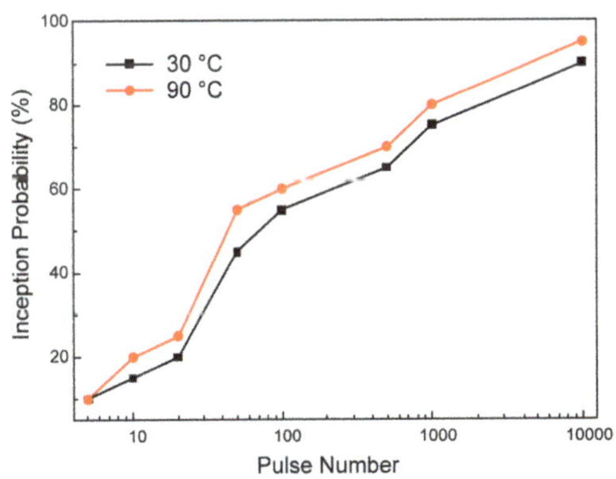

$$E = \frac{2U}{R\ln\left(\frac{1+4d}{R}\right)} \quad (2\text{-}1)$$

Where E characterizes the electric field at the tip of the needle, U characterizes the pulse amplitude, R characterizes the radius of curvature at the tip, and d characterizes the gap between the needle and the planar electrode. (I llias,

Figure 10. Relationship between the breakdown probability and the temperature

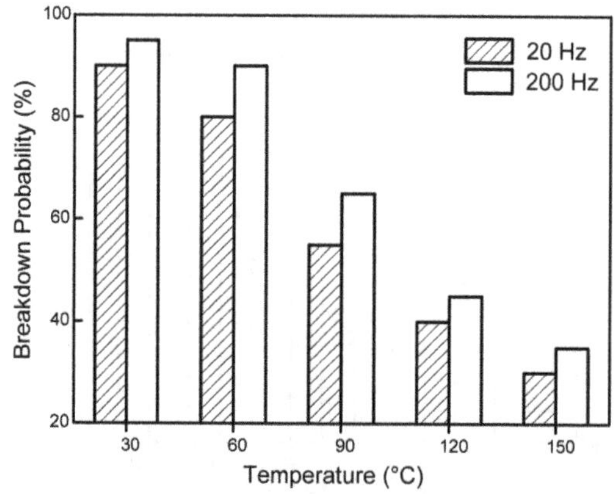

2015) reported the rise time of partial discharge in the pulse voltage, similar to that reported by (Luo, 2014). The minimum electric field of partial discharge in silicone rubber is E min. The delay effect of the initial electron should be considered on the first pulse waveform. When E - Δ E > E min, partial discharge and residual electric field in the channel tree into E-res. After the first partial discharge, the electric field E will revert to E min for a given tree channel. In Figure 11, a new partial discharge occurs. Partial discharge number n-pulse is determined by the following formulation:

$$\mathrm{n}_{pulse} = K\frac{dE}{dt}\frac{t_d}{(\mathrm{E}_{\min} - E_{res})} \qquad (2\text{-}2)$$

When the pulse amplitude rises, td1>td, $\frac{dE_1}{dt} > \frac{dE}{dt}$ and n-pulse becomes larger. The hot electron formed by partial discharge could gain higher energy at higher pulse amplitude to violent collide with the molecular chain.

After partial discharge in tree channel, charge Q can accumulate near the tip and surface of the tree channel. The electric charge in the wall of the tree channel can form internal electric field Ed and eventually produce partial discharge.

The charge Q will spread in the polymer and decrease the internal Ed. Dissipation rate associated with tau d, tau d is determined by the tree dimensions d and dielectric constant epsilon. Then calculate Ed using formulas (3) and (4):

Figure 11. Endured electric field of tree channel at different pulse amplitude

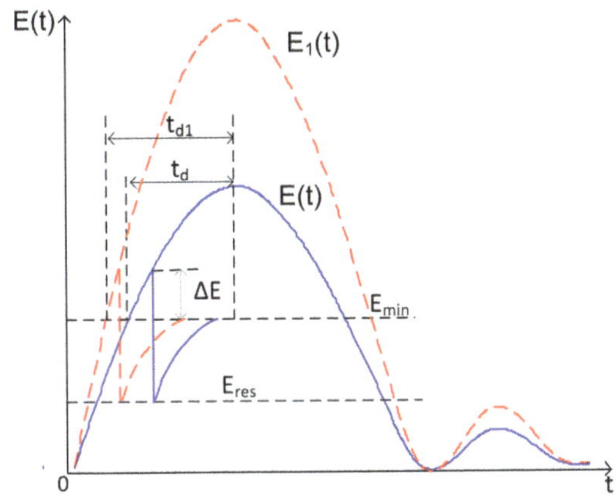

$$\tau_d = \varepsilon D / \delta_s \tag{2-3}$$

$$E_d = Aq(\mathrm{t}+\Delta\,\mathrm{t}) = A\,\mathrm{q(t)}\exp(-\frac{\Delta t}{\tau_d}) \tag{2-4}$$

The constant related to the charge transfer rate, Δ t is the time interlude between two pulses which related to tree channel size. When the pulse frequency rises from 5 to 1000Hz, Ed (5Hz) <Ed (1000Hz). Assuming that Emin remains constant for a tree channel, when Ed covers the next pulse waveform E (t), the effective time tac of partial discharge was greater than tbc. According to formula (2), n-pulse increases with the increase of the pulse frequency, which can be explained by the Figure 12. Local high temperature is produced by partial discharge. Local high temperature could soften the surface tree channel and form thermal damage to the matrix. The thermal does not dissipate soon during the short interval between the two pulses. So, the thermal damage accumulates after partial discharge. Considering the above factors, as shown in Figure 8, with an increase in pulse frequency, the initial probability of electric tree becomes larger.

In (Shimizu, 1996), it has been reported that free volume played an significant role in the formation of electrical trees. In free-volume area, the injected electrons are enhanced to collide with the molecular chain of polymer.

Figure 12. Effective time of partial discharge with different pulse frequency

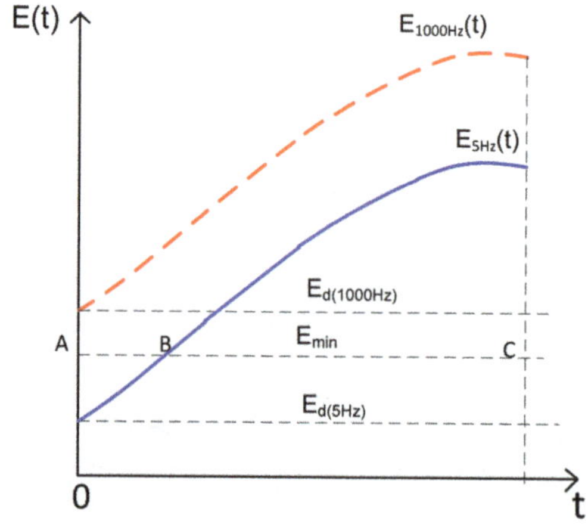

Therefore, with the increase of temperature environment, the electrical and mechanical properties of the polymer matrix would also change (Li, 2014).

In (Artbauer, 1996), the heat absorption peak of XLPE is about 120°C in DSC curve. When Tamb rises from 40°C to 80°C, XLPE changes from a glassy state to a highly elastic state. The growth of free trading volume depends on (li, 2014):

$$f_V = f_g + \Delta\alpha(\mathrm{T}-\mathrm{T}_g)\,\mathrm{T}>\mathrm{Tg} \tag{2-5}$$

FV said scores () associated with the total quantity of free volume, fg said free volume, under temperature Tg Δ said temperature Tg cubic the differences between the thermal expansion coefficient value above and below. According to the theory of free volume, electrical failure is related to the longest free path lx. Decomposition criteria can be illuminated by the following equation (Artbauer, 1996):

$$\mathrm{W}=\mathrm{e}E_b l_x \tag{2-6}$$

Where W is the obtained electron energy in lx, Eb is the stress electric field and serious deformation occurs around the tip of the needle and the tip of the channel, and e is the electron charge quantity. As the temperature increases, the fV gets bigger, and the lx gets stretched. Thus, ionization through electron collisions is enhanced over longer paths, which speeds up the breaking of molecular chains. When Tamb is 50°C, the electric tree propagates in crystal spherulites and generates shrubby structures of 16 or 19 kV.

However, with an increase in temperature, the molecular chain thermal motion becomes energetic and the relaxation time of the chain segment increases. The free volume expands at higher temperatures and space for the chain to move becomes larger. In this way, the effect physical connections between molecule is decreased and intermolecular forces are reduced. The tensile strength, tear strength and elastic modulus of silicone rubber are reduced. These factors lead to the decrease of the mechanical property of silicone rubber at high temperature. Therefore, the initial energy threshold of the tree is lowered.

Electrons can gain energy in their free volume and then collide with one another, creating high energy electrons and break the molecule chain. During this process, local high pressure and high temperature will be formed, which will cause thermal as well as mechanical damage to the sample. At high temperatures, as the free volume expands, the heat electron energy gained increases. Some injected charge was reported to be trapped in the matrix. The trapped charge around the dense bush can generate a shield around the bush channel tops, which weaken the electric field. In this case, the tree length is shorter for N-pulse = 10000, and longer even in the initial phase as can be seen in Figure 6a.

Figure 13. Tensile strength and tear strength as a function of temperature

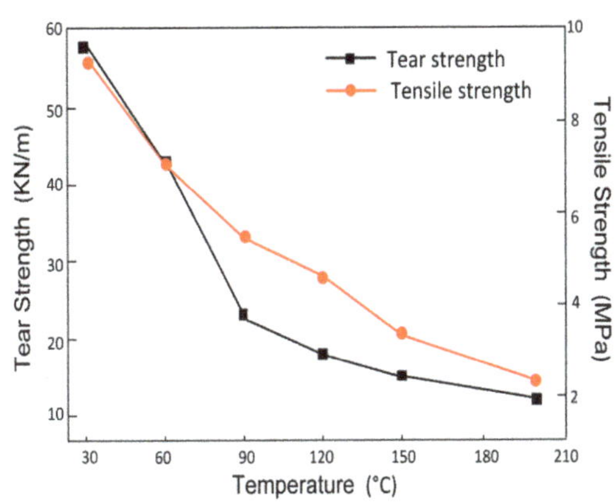

EFFECTS OF MECHANICAL STRESS ON TREEING GROWTH CHARACTERISTICS IN HTV SILICONE RUBBER

The form of electric tree can be described based on the number of tree channels in the process of trearization (Du, 2015). Typically, they can be divided into shrubs, branches, pine, bark and mixed structural structures (Ahmad, 2004). The difference of electrical tree morphology depends on the existence of discharge explosion. The long discharge path formed in a large number of discharge activities could reach the treetops and is conducive to the extension of tree channels, which is conducive to the formation of branching trees. Considering both the tensile stress and compressive stress, the propagation and morphological distribution trend of electric trees were obviously observed after the treeing time about 600 s. After 600 s, the shrub remained in stasis for a long time until a rapid collapse occurred. Table 2 shows the typical electrical tree morphology with different tensile rates. When the impulse voltage is 6 kV and the tensile ratio is 0% or 10% respectively, the branches and bushes exist at the same time. But when the impulse voltage is 12 kV, they are shrubs regardless of the elongation. Considering the tensile stress and compressive stress, the propagation and morphological distribution trend of the electric trees was obvious after the trearization time of 600s. After 600s, the bushes remained stagnant for a long time until rapid collapse occurred, as shown in table 2. Typical electrical tree morphology at different tensile rates. When the impulse voltage is 9 kV and the tensile ratio is 0% or 10% respectively, the branches and bushes exist simultaneously. When the pulse voltage is 12 kV, they are shown in table 3, which shows the appearance of the typical electrical tree form at different compression rates. The different distribution of electrical tree morphology is related to the microstructural changes of silicone rubber under mechanical stress, which will be discussed in the next section.

Figure 14 shows the probability distribution of typical electric tree morphology under different 9kv mechanical stress. As elongation increases from 0% to 30%, the probability of branching trees increases from 60% to 0%,

Table 2. Appearance of typical electrical tree morphologies at different tensile rates

	0	10%	20%	30%
9 kV	bush/branch tree	bush/branch tree	bush tree	bush tree
12 kV	bush tree	bush tree	bush tree	bush tree

Table 3. Appearance of typical electrical tree morphologies at different compressive rates

	0	10%	30%	50%
9 kV	bush/branch tree	bush/branch tree	branch tree	branch tree
12 kV	bush tree	bush/branch tree	bush/branch tree	branch tree

Figure 14. Appearance probability of typical electrical tree morphologies at different mechanical stress at 9 kV

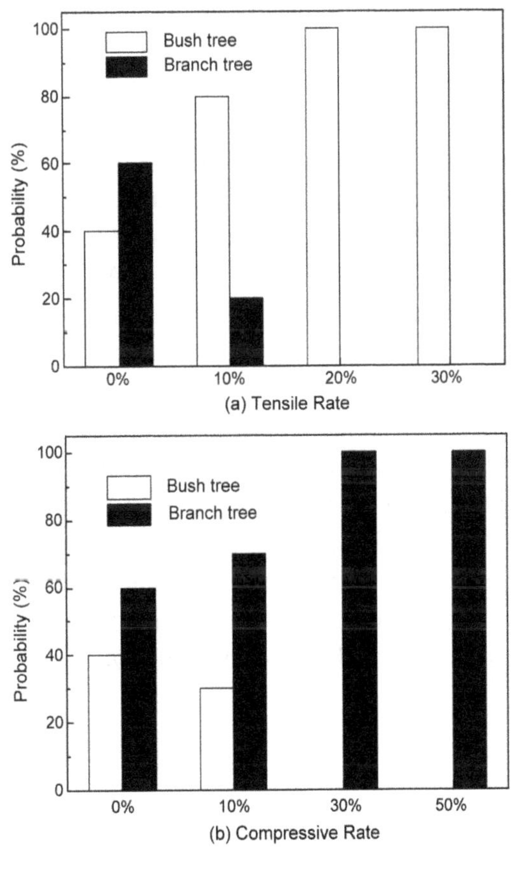

and the probability of shrubs increases from 40% to 100%.As compression rates increased from 0% to 50%, the incidence of branching trees increased from 60% to 100%, and shrubs from 40% to 0%. As elongation increased from 0% to 30%, the probability of branching trees increased from 60% to 0%, and the probability of shrubs increased from 40% to 100%.As compression rates increased from 0% to 50%, the incidence of branching trees increased from 60% to 100%, and shrubs from 40% to 0%.This shows that the morphology of the electric branches is greatly affected by tensile and compressive stresses.

Figure 15 shows the typical morphology of electric trees at different tensile rates after 600 s. At a higher tensile rate, the tree presents denser structures and larger failure areas, which indicates that tensile stress accelerates the growth of the tree. The initiation and growth of electric branches are closely related to the movement of electric charge in materials. High-energy electrons generated for various reasons will bump into and destroy the molecular chain of insulating materials and gradually form a low-density area, which will eventually lead to partial discharge and form electric branches. Stress can actually affect the free volume distribution of the insulating material and thus the free travel of the charge in the material and thus the electrical branch ability of the material, which will be discussed in detail in the following sections.

After about 600 s of tree growth, Figure 16 shows (a) changes in tree length and (b) changes in fractal size with stretching rate. At 9kV, the average length of trees is about 0.5-0.9mm. When the elongation rate increases from 0% to 30%, it gradually increases from 1.0mm to 1.2mm at 12kV.These results show that at a higher tensile rate, the electric tree is more likely to propagate in the direction of the electric field. As the stretch ratio increased from 0% to 30%, the average fractal dimension increased from 1.6 to 1.75 at 9kV and from 1.73 to 1.93 at 12kV.The results show that the structures of electric trees are more complex and diverse at higher tensile rates.

Cumulative damage is to extract the total number of pixels covering the tree area, which is often used to describe the damaged area of the tree. Figure 17 shows the relationship between cumulative damage and bifurcation

Figure 15. Typical morphologies of electrical trees at different tensile rates

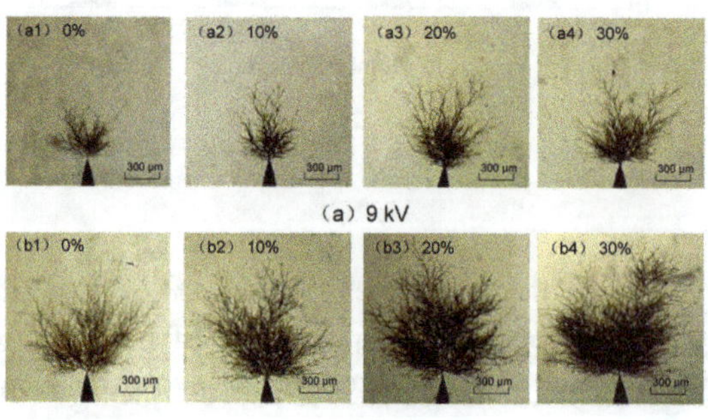

Figure 16. (a) Tree length and (b) fractal dimension as a function of tensile rate

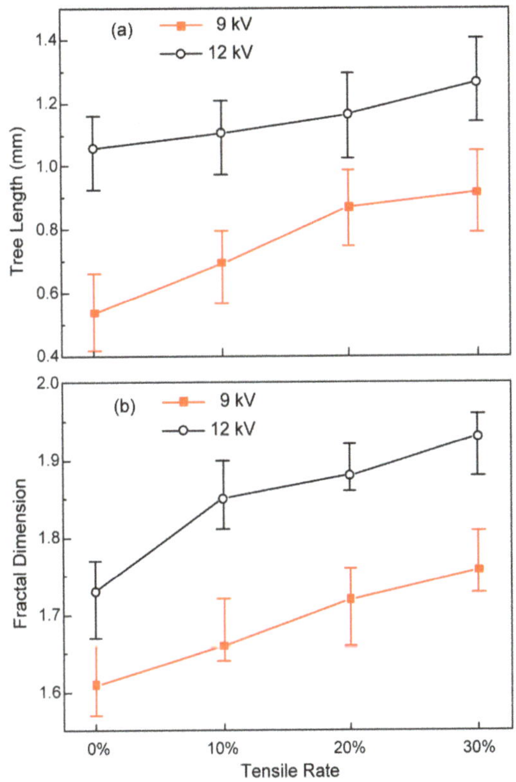

time at different tensile rates. When the bifurcation time reached 600 s and the tensile rate increased from 0% to 30%, the cumulative damage at 9kV increased from 7×103 pixels to 12×103 pixels and from 14×103 pixels to 103×40×103 pixels. The results show that the area of the tree is larger at a higher stretching rate. In addition, it should be noted that the cumulative damage of specimens without tensile stress is less than that of specimens with tensile stress.

Figure 18 shows the typical form of an electric tree at different compression rates. With the increase of compression rate, at 12kV, the morphology of electric trees between thickets will change obviously. With the increase of compression rate, the number of electrical tree channels decreased obviously. The initiation and growth of electric branches are closely related to the movement of electric charge in materials. High-energy electrons generated for various reasons will bump into and destroy the molecular chain of insulating

Figure 17. Relationship between the accumulated damage and the treeing time with different tensile rates

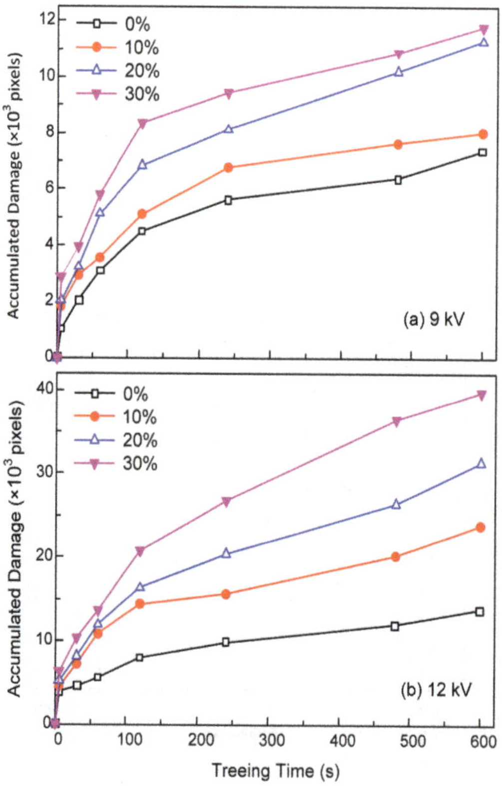

materials and gradually form a low-density area, which will eventually lead to partial discharge and form electric branches. Stress can actually affect the free volume distribution of the insulating material and thus the free travel of the charge in the material and thus the electrical branch ability of the material, which will be discussed in detail in the following sections.

Figure 19 shows the relationship between (a) branch length and (b) fractal dimension and compression ratio after 600s. As can be seen from the figure, the average value of tree length and fractal dimension decreases with the increase of compression ratio. The results show that the proposed algorithm can effectively suppress the development of the electric tree in the direction of electric field and generate a simple branch structure with high compression ratio.

Figure 18. Typical morphologies of electrical trees at different compressive rates

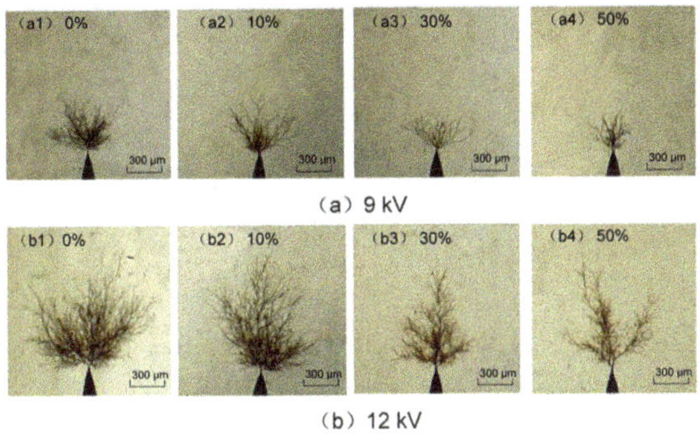

As shown in Figure 6, with the increase of compression ratio, the electrical morphology of trees changed from shrub to branch tree, and the cumulative damage gradually decreased. Figure 20 shows the relationship between cumulative damage and branching time at different compression rates. After about 600s, the cumulative damage under 9kv increased from 7×103 pixels to 2.2×103 pixels, and from 14×103 pixels to 14×103 pixels. When the compression ratio increased from 0% to 50%, the pixels became 6×103 pixels (12 kV). The cumulative damage of specimens without tensile stress is greater than that of specimens with compressive stress. The results are contrary to those under tensile stress described in section 3.2. These are discussed in detail in a later section.

Silicone rubber is an inorganic polymer material, including the main chain of the si-o inorganic bond and the side chain of the ch3 organic bond. Silicone rubber is an amorphous polymer with irregular molecular chains. It is in an amorphous state at room temperature, so it is easy to deform under the action of mechanical stress. The reduction of mechanical properties affects the arrangement of intermolecular attractive bonds (Crine, 2005). Moreover, metamorphism may even affect homolysis of valence bonds, in which atoms join together to form macromolecules. The molecular chain of silicone rubber will be rearranged under the action of stress, because the stress will affect the binding energy of the molecular chain, this process will change the free volume distribution of polymer, and then affect the electrical branch ability of silicone rubber. The initiation and growth of electric branches are closely related to the movement of electric charge in materials. High-energy electrons generated for various reasons will bump into and destroy the molecular chain of insulating materials and gradually form a low-density area, which will

Figure 19. (a) Tree length and (b) fractal dimension as a function of compressive rate

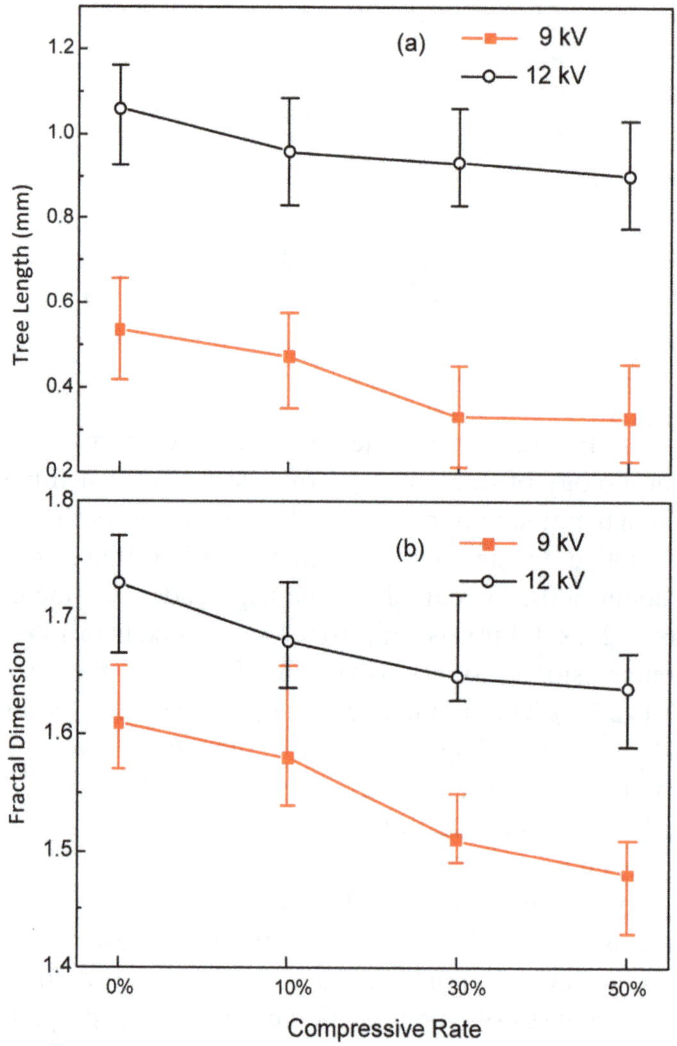

eventually lead to partial discharge and form electric branches. Stress can actually affect the free volume distribution of the insulating material and thus the free travel of the charge in the material and thus the electrical branch ability of the material, which will be discussed in detail in the following sections.

When stress is applied to the silicone rubber, the physical structure and chemical bonds on the chain are broken. As a result, the molecular chain

Figure 20. Relationship between the accumulated damage and the treeing time with different compressive rates

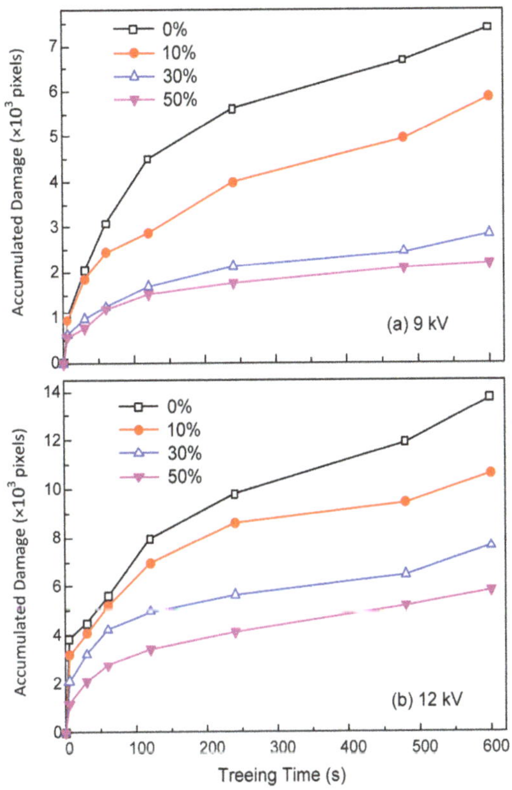

structure is also damaged, which can lead to more physical defects and accelerate the tree development process. The free volume was determined by the positron invisible spectrum (PALS).When silicone rubber is under compression stress, with the increase of physical connection points, the distance between molecular chains gradually decreases, forming more attractive chemical bonds, which is conducive to the stability of the structure of silicone rubber. In addition, most microcavities are compressed in the direction of compressive stress. Free volume goes down. The partial discharge behavior of silicone rubber is restricted and the dendrite growth of silicone rubber is hindered. This indicates that the stress actually affects the electrical branch ability of the material by affecting the free volume distribution of the material, and the free volume is closely related to the electrical branch ability.

Figure 21. Schematic diagram of microscopic structure of silicone rubber under (a) tensile stress and (b) compressive stress

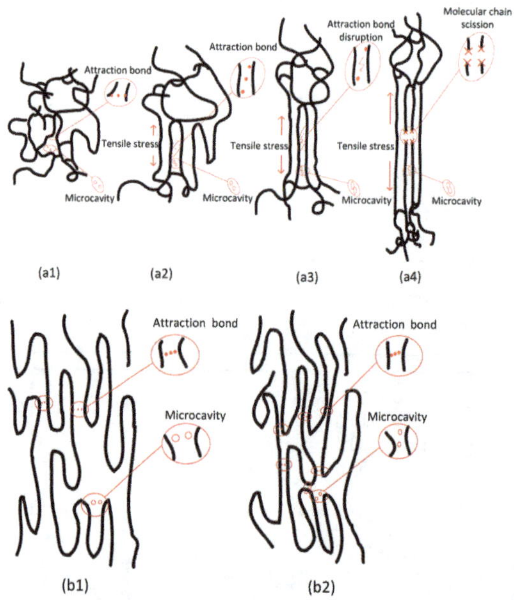

Decomposition equation can be explained by the following formula (d, 2005; Ding, 1994):

$$W=eE_b l_x \tag{2-7}$$

Where W is the electron energy gained in microcavity and Eb is stress electric field in it and serious deformation occurs around the tip of the needle and the tip of the tree channel, e is the electron charge and the free path of the longest electron in lx. When the sample is subjected to tensile stress, the microcavity is suggested to be stretched along the tensile stress direction. As shown in Figure 22, the free path along the Ed direction in the microcavity tends to be longer, and the longest free path of 1x is improved. Electrons can be accelerated and gain more energy in the stretched chamber.

The dendrite process can generally be divided into three stages, including the initial stage, the expansion stage and the escape stage, but the initial stage may vary according to the electrical and mechanical stresses, starting with pores, impurities, bumps or mechanical deformation. High energy charge

Figure 22. Free path of electrons in the microcavity

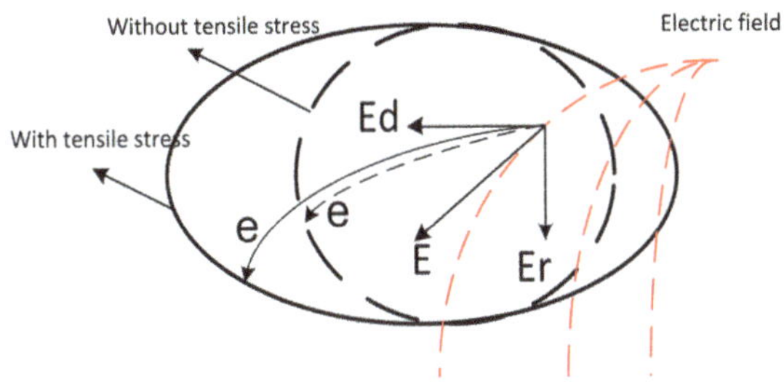

injection can cause mechanical, thermal and chemical damage to the polymer. Therefore, the area of the polymer tip should be determined according to the damage process caused by the injected charge before observing the tubular tree channel. Figures 23a and 23 are the first phases of Figure 23. As shown in Figure 24, a low pulse voltage applied to the tip does not immediately generate an electric tree. When the deformation electric field is large enough, partial discharge occurs and a large number of hot electrons collide with the molecular chain. Generally speaking, it is believed that thc formation of electrical branches is closely related to high-energy electrons, which will break the molecular chain of the polymer, gradually form a low-density area, and eventually trigger partial discharge, forming macroscopic observable electrical branches. The interaction between thermionic and polymer structures is the main mechanism of tree initiation (Shimizu, 1998).When the kinetic energy generated by the hot electrons is sufficient to cause chemical degradation,

Figure 23. Typical tree morphologies during the electrical tree initial process at 6 kV

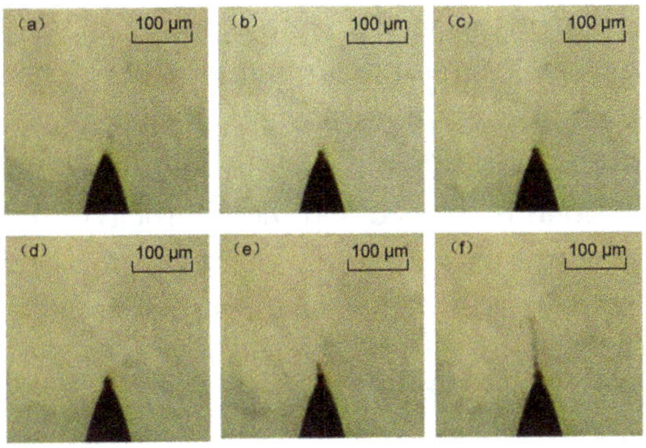

Figure 24. Relationship between the tree length and treeing time during the tree initial process at 6 kV

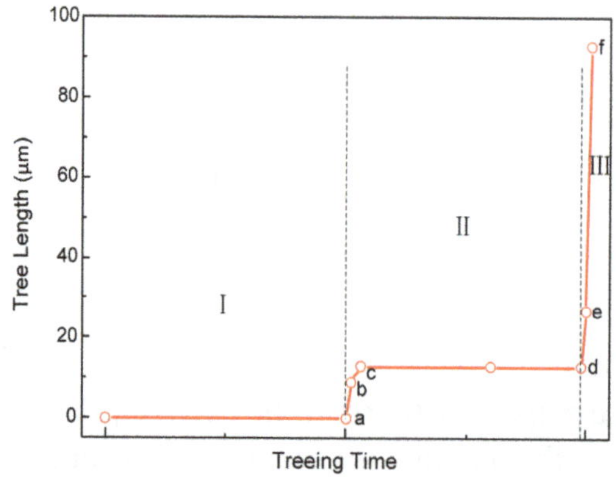

the electric tree structure is formed around the tip of the molecular chain, as shown in Figures 23b, 23c, 23d and 242 platforms.

As described in the previous section, the high temperature combination of hot electron collisions causes the main chain to break into small molecules, while the volatile component of the volatile small molecule, volatile cyclosiloxane, accumulates and causes an increase in pressure. When the yield strength of silicone rubber is greater than that of silicone rubber, new pores can form, some of which may break and even form a string of pearls. In addition, the electric field in the channel causes severe deformation of the blade tip, which greatly increases partial discharge and accelerates two factors of rapid growth of trees. Therefore, mechanical stress will greatly affect the electrical branch characteristics of silicone rubber.

When silicone rubber is subjected to mechanical stress, the microstructure of the molecular chain will change significantly, as shown in Figure 21. Under the action of tensile stress, the gravitation between molecular chains decreases, and the microcavity is stretched, causing physical deformation. Defects and increased free volume. Under the action of compression stress, the more attractive bond is formed and the microcavity is suppressed. The change of microstructure will affect the behavior of partial discharge, and partial discharge has a great influence on the initiation and growth of electric tree. The initial mechanism of the electric tree considering the effect of mechanical stress is shown in Figure 25.

Figure 25. Initial mechanism of electrical tree

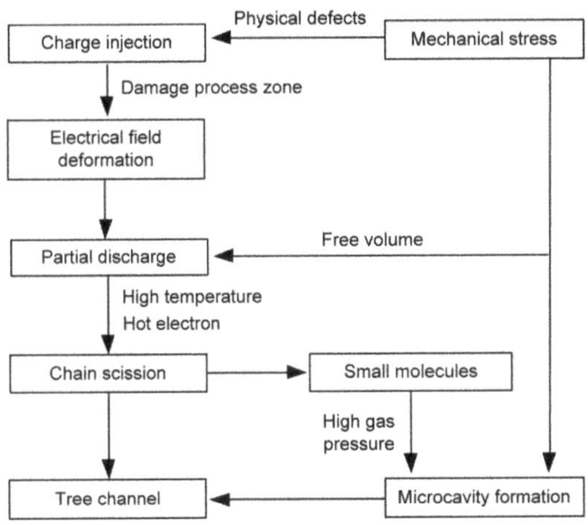

To gain insight into the relationship between the initial time and mechanical stress of an electrical trcc, Figurc 27 shows the initial time of an electrical tree with different (a) tensile rates and (b) compressibility. At 6kv, the average initial time varies from 25 seconds to 10 seconds as the tensile rate increases from 0% to 30%.When the compression ratio increased from 0% to 50% at 6kv, the average initial time of the electrical tree increased from 25 seconds to 100 seconds, contrary to the trend under tensile stress. The initiation and growth characteristics of electrical branches are consistent, that is, tensile stress will accelerate the formation and development of electrical branches, while compressive stress will inhibit the formation and development of electrical branches. This is due to the different effects of different stresses on the microstructure of the materials, so the effects on the properties of the electrical branches are also different.

CONCLUSION

Different temperature, pulse frequency and mechanical stress will affect the electrical tree properties of Sir. When N pulse = 10000, although Tmb increased to 150°C, only shrub shape was observed, although the pulse amplitude was larger. Speed up tree growth by emphasizing higher pulse frequencies. After the same number of pulses, the length of the tree and FD

Figure 26. Initiation probability of electrical tree with different (a) tensile rates and (b) compressive rates

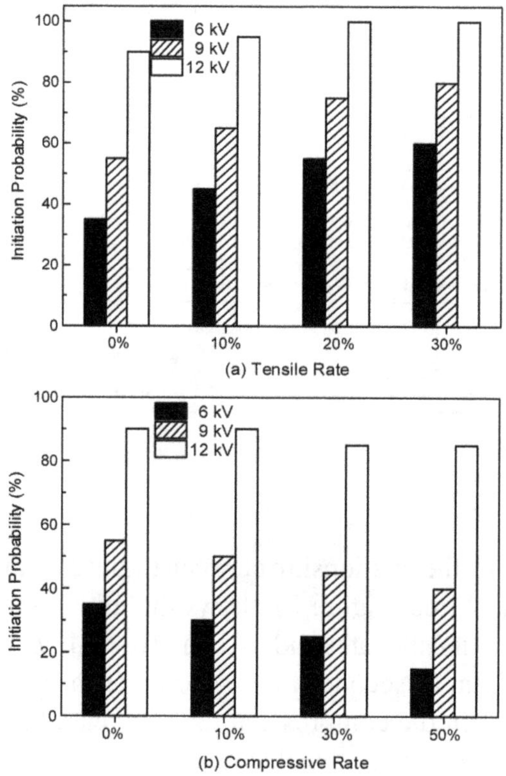

both tend to increase with the increase of pulse frequency. Tree initiation and breakdown characteristics are closely related to high temperature and pulse frequency. The initial probability of the tree increases with the increase of pulse frequency and number of pulses. The shape of the tree is related to the mechanical stress of the specimen. Both tensile stress and compressive stress affect the distribution of electric tree morphology. Under different mechanical stress, the microstructure of silicone rubber will change, which will affect the partial discharge characteristics of the sample and the propagation characteristics of the tree. Under the action of tensile stress, with the increase of tensile rate, the length, fractal dimension and cumulative damage of trees all increased, indicating that higher tensile rate would accelerate the growth process of trees. The initial mechanism of electrical tree is related to the mechanical properties of silicone rubber. Under the action of tensile stress,

Figure 27. Initial time of electrical tree with different (a) tensile rates and (b) compressive rates

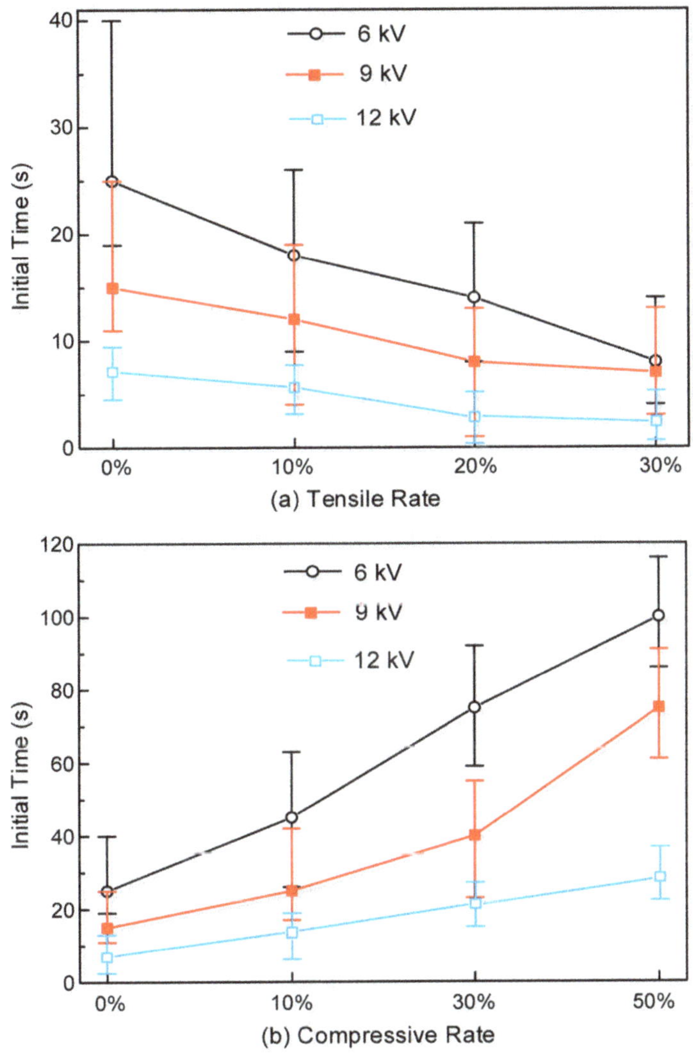

the gravitation between the molecular chains is weakened, and the microcavity is stretched or even fused into a large microcavity. The initiation and growth of electrical branches are closely related to the free volume of materials. Different mechanical stresses will have different effects on the molecular chain arrangement of materials, which will further have different effects on the characteristics of electrical tree.

REFERENCES

Ahmad, M. H., Bashir, N., Ahmad, H., Abd Jamil, A. A. & Suleiman, A. A. (n.d.). An Overview of Electrical Tree Growth in Solid Insulating Material with Emphasis of Influencing Factors, Mathematical Models and Tree Suppression. *TELKOMNIKA Indonesia J. Electr. Eng.*, *12*(8), 5827-5846.

Al-Ghamdi, S. A., & Varlow, B. R. (2004). Treeing in Mechanically Pre-stressed Electrical Insulation. *IEEE Transactions on Dielectrics and Electrical Insulation*, *11*(4), 130–135. doi:10.1109/TDEI.2004.1266326

Artbauer, J. (1996). Electric Strength of Polymers. *Journal of Physics. D, Applied Physics*, *29*(1), 446–456. doi:10.1088/0022-3727/29/2/024

Billing, J. W., & Groves, D. J. (1974). Treeing in Mechanically Stressed HV Cable Polymers Using Conducting Polymer Electrodes. *Proc. IEE*, *121*(11), 1451-1456.

Chen, G., & Tham, C. H. (2009). Electrical Treeing Characteristics in XLPE Power Cable Insulation in Frequency Range Between 20 and 500 Hz. *IEEE Transactions on Dielectrics and Electrical Insulation*, *16*(1), 179–188. doi:10.1109/TDEI.2009.4784566

Chen, X. R., Xu, Y., Cao, X. L., & Gubanski, S. M. (2015). Electrical Treeing Behavior at High Temperature in XLPE Cable Insulation Samples. *IEEE Transactions on Dielectrics and Electrical Insulation*, *22*(5), 2841–2851. doi:10.1109/TDEI.2015.004784

Crine, J. P. (2005). Influence of Electro-mechanical Stress on Electrical Properties of Dielectric Polymers. *IEEE Transactions on Dielectrics and Electrical Insulation*, *12*(4), 791–800. doi:10.1109/TDEI.2005.1511104

Ding, H. Z. & Varlow, B. R. (2005). Thermodynamic Model for Electrical Tree Propagation Kinetics in Combined Electrical and Mechanical Stresses. *IEEE Transactions on Dielectrics and Electrical Insulation.*, *12*(1), 81-89.

Ding, H. Z., Xing, X. S., & Zhu, H. S. (1994). A Kinetic Model of Time-dependent Dielectric Breakdown for Polymers. *Journal of Physics. D, Applied Physics*, *27*(1), 591–595.

Dissado, L. A. (2002). Understanding Electrical Trees in Solids from Experiment to Theory. *IEEE Transactions on Dielectrics and Electrical Insulation*, *9*(4), 483–497. doi:10.1109/TDEI.2002.1024425

Dissado, L. A. (2002). Understanding Electrical Trees in Solids from Experiment to Theory. *IEEE Transactions on Dielectrics and Electrical Insulation*, *9*(4), 2002. doi:10.1109/TDEI.2002.1024425

Du, B. X., Su, J. G., & Han, T. (2017). Temperature-dependent electrical tree in silicone rubber under repetitive pulse voltage. *IEEE Transactions on Dielectrics and Electrical Insulation*, *24*(4), 2291–2298. doi:10.1109/TDEI.2017.006461

Du, B. X., Su, J. G., Li, J., & Han, T. (2017). Effects of mechanical stress on treeing growth characteristics in HTV silicone rubber. *IEEE Transactions on Dielectrics and Electrical Insulation*, *24*(3), 1547–1556. doi:10.1109/TDEI.2017.006123

Eichhorn, R. M. (1977). Treeing in Solid Extruded Electrical Insulation. *IEEE Transactions on Dielectrics and Electrical Insulation*, *12*(1), 2–18. doi:10.1109/TEI.1977.298001

Illias, H. A., Tunio, M. A., Mokhlis, H., Chen, G., & Bakar, A. H. A. (2015). Experiment and Modeling of Void Discharges Within Dielectric Insulation Material Under Impulse Voltage. *IEEE Transactions on Dielectrics and Electrical Insulation*, *22*(4), 2252–2260. doi:10.1109/TDEI.2015.004817

Kudo, K. (1998). Fractal Analysis of Electrical Trees. *IEEE Transactions on Dielectrics and Electrical Insulation*, *5*(5), 713–727. doi:10.1109/94.729694

Li, S. T., Wang, W. S., Yu, H., & Sun, H. G. (2014). Influence of Hydrostatic Pressure on Dielectric Properties of Polyethylene/Aluminum Oxide Nanocomposites. *IEEE Transactions on Dielectrics and Electrical Insulation*, *21*(21), 519–528. doi:10.1109/TDEI.2013.004131

Luo, Y., Wu, G. N., Liu, J. W., Zhu, G. Y., Wang, P., Peng, J., & Cao, K. J. (2014). PD Characteristics and Microscopic Analysis of Polyimide Film Used as Turn Insulation in Inverter-fed Motor. *IEEE Transactions on Dielectrics and Electrical Insulation*, *21*(5), 2237–2244. doi:10.1109/TDEI.2014.003868

Mason, J. H. (1955). Breakdown of Solid Dielectrics in Divergent Fields. *Proc. IEE-Part C: Monographs*, *102*(2), 254-263.

Nakanishi, K., Hirabayashi, S., & Inuishi, Y. (1979). Phenomena and Mechanisms of Tree Inception in Epoxy Resins. *IEEE Transactions on Dielectrics and Electrical Insulation*, *14*(6), 306–314. doi:10.1109/TEI.1979.298186

Shimizu, N., & Laurent, C. (1998). Electrical Tree Initiation. *IEEE Transactions on Dielectrics and Electrical Insulation*, *5*(5), 651–659. doi:10.1109/94.729688

Shimizu, N., & Sato, H. (1996). Process in PE Free-volume Preceding Electrical Tree Initiation. *IEEE Conf. Electr. Insul. Dielect. Phenom. (CEIDP)*, *2*(1), 787-790. 10.1109/CEIDP.1996.564626

Varlow, B. R. (2002). Mechanical Prestressing Improves Electrical Strength. *IEEE Transactions on Dielectrics and Electrical Insulation*, *18*(1), 12–25. doi:10.1109/57.981323

Varlow, B. R., & Malkin, G. J. (2000). Electrical Treeing in Mechanically Prestressed Insulation. *IEEE Transactions on Dielectrics and Electrical Insulation*, *7*(6), 721–724. doi:10.1109/94.891981

Wu, K., & Dissado, L. A. (2005). Model for Electrical Tree Initiation in Epoxy Resin. *IEEE Transactions on Dielectrics and Electrical Insulation*, *12*(4), 655–668. doi:10.1109/TDEI.2005.1511091

Chapter 4
Discharge and Flashover Behavior in Oil-Paper

ABSTRACT

Due to special operating conditions, the valve side bushing of the converter transformer connected to the converter valve is subject to complex voltage excitation, including DC voltage, AC/DC composite voltage, lightning impulse overvoltage, or composite voltage of operating overvoltage and DC. Under the action of this complicated electric field, the oil-paper insulation of the valve-side bushing of the converter transformer is prone to electric field distortion due to charge accumulation, which causes a surface discharge, which will seriously cause the edge breakdown. At the same time, since the temperature in the converter transformer rises due to a large amount of loss during the operation of the transformer, creeping discharge is more likely to occur under the electrothermal composite field. Hence, it is significant to carry out research on the surface discharge characteristics of the oil-paper insulation on the valve side of the converter transformer under the electrothermal composite field.

DOI: 10.4018/978-1-5225-8885-6.ch004

INTRODUCTION

UHVDC transmission has the advantages of large capacity and low loss. Converter station equipment is the most critical part of HVDC technology, and its reliability is the key to ensure the safe and stable operation of the whole system. (Flourentzou, 2009; Piovan, 2010). Oil-paper insulation is one of the most commonly used insulating materials in converter transformers due to its excellent electrical and mechanical properties and good thermal stability. Due to the complex operating environment of the converter transformer, the oil-paper insulation in the side winding of the valve should bear not only DC voltage and AC voltage, but also various overvoltages (Chao, 2010; Wang, 2012; Sima, 2015). In actual operation, the overvoltage usually consists of the pulse voltage superimposed on the working DC voltage. In addition, the temperature rise in the converter transformer will also lead to the decline of the reliability of oil-paper insulation (Vasa, 2017).

Under the action of electric field for a long time, electric charge tends to accumulate on the dielectric surface, which will cause local electric field distortion and then cause surface flashover and insulation system failure (Leblanc, 2015; Jadidian, 2012). In addition, the surface charge behavior of the insulation system becomes more complicated under the superimposed voltage, and the surface flashover of the dielectric is more likely to occur under the superimposed pulse voltage of the DC voltage. It has been proved that the frequency, amplitude, number and polarity of the pulse voltage all affect the accumulation and dissipation of surface charges (Du, 2015a, 2015b). The dielectric breakdown characteristics of oil-paper insulation system depend on the polarity combination of pulse voltage and DC voltage (Okabe, 2015; Du, 2018).

In addition, temperature will speed up the process of the SPD, which affect the dielectric distributions of traps (Zhou, 2017; Du, 2017). The surface flashover of the dielectric is also closely related to temperature. It is found that the initial voltage of partial discharge decreases with the increase of temperature (Zhou, 2013). When polarity reversal voltage exists, the decrease of accumulated charge on the dielectric surface will cause the flashover voltage to rise (Du, 2016). However, there are few studies on the surface charge and flashover properties of oil-paper insulation under the action of electric and thermal fields. Therefore, this chapter aims to explore the flashover mechanism of oil-paper insulation, which is affected by superimposed voltage and temperature.

In this chapter, the surface charge of oil-paper insulation is studied by using SPD measurement. Most importantly, this chapter discusses the mechanism of electric heating compound field affecting flashover. According to this study, the surface charge and flashover characteristics of oil-paper insulation under electrothermal composite field can be more clearly understood, which will provide a basis for eliminating surface flashover faults in practice.

EFFECT OF SUPERIMPOSED VOLTAGE ON SURFACE CHARGE AND FLASHOVER

Effect of Superimposed Voltage on Surface Charge

The corona discharge occurs after the DC voltage is applied to the tip. Under the action of the electric field, the applied voltage generates charges of the same polarity, which is injected into the surface of the sample. The accumulated charge is mainly deposited in a thin layer of micron below the upper surface of the sample, which in turn changes the surface potential of the sample. Based on the results of the SPD measurement, the initial surface potential and the decay rate were obtained (Du, 2016). Figure 1 and 2 shows the initial surface potential and attenuation rate of the DC voltage superimposed at different pulse voltages. The effect of the pulse voltage on the accumulation of surface charge can be seen more intuitively as the initial surface potential changes with the pulse voltage. When the superimposed pulse voltage is the same as the DC voltage polarity, the initial surface potential increases as the pulse voltage increases, as shown in figures 2. When the superimposed pulse voltage is different from the DC voltage polarity, as the pulse voltage is large, the initial surface potential first increases and then decreases, as shown in figure 1. According to the variation curve of the surface potential decay rate in figure 1 and 2, it can be seen that the surface potential decay rate change is the same as the initial surface potential.

Figure 3 shows charge injection when DC voltage is preapplied at 5kV and negative polarity pulse voltage is superimposed. Figure 3 (a) shows the situation when the pulse voltage is less than the DC voltage. In this section, the pulse voltage is -3kV, and the composite voltage is 2kV, which is still greater than 0. When the pulse voltage rises to -5kV, that is, the composite voltage

Figure 1. Changes of initial surface potential and decay rate under superimposed voltages with different polarities at 20 °C

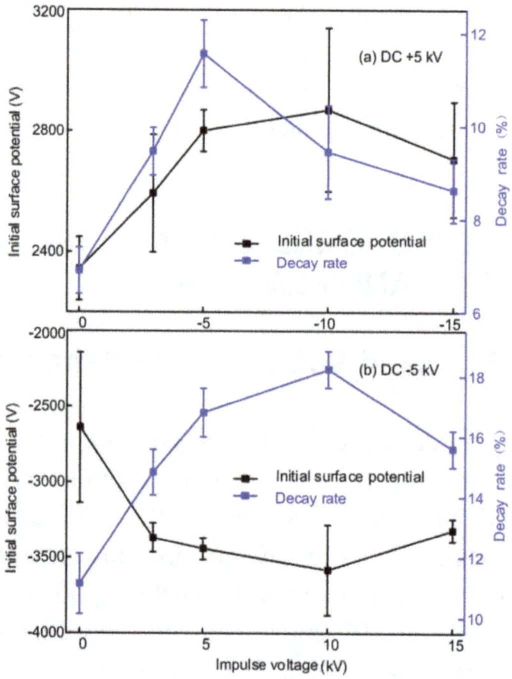

is 0, as shown in Figure 3 (b). At this time, because the composite voltage did not cross zero, there was still no negative charge injection. However, at the rising edge of the composite voltage, in the same time as the rising edge in figure 3 (a), the voltage rose from 0 to 5kV rapidly, during which more positive charge was ionized than (a), so the initial surface potential continued to increase. When pulse voltage up to 10 kV, compound voltage of 5 kV, at this point in the pulse voltage rise along the injection will have a negative charge, but fewer composite with positive charge on the surface of the sample, in compound voltage rise along the voltage rise faster ionization out more positive charge, is made up for with negative charge injection and compound on the surface of positive charge, also makes the surface potential is higher than that of pulse voltage is 5 kV surface potential.

However, as the pulse voltage continues to increase to -15kV, the amount of negative charge injected by the rising edge of the pulse is more, which is compounded with more positive charge, so that the surface potential decreases

Figure 2. Changes of initial surface potential and decay rate under superimposed voltages with the same polarity at 20 °C

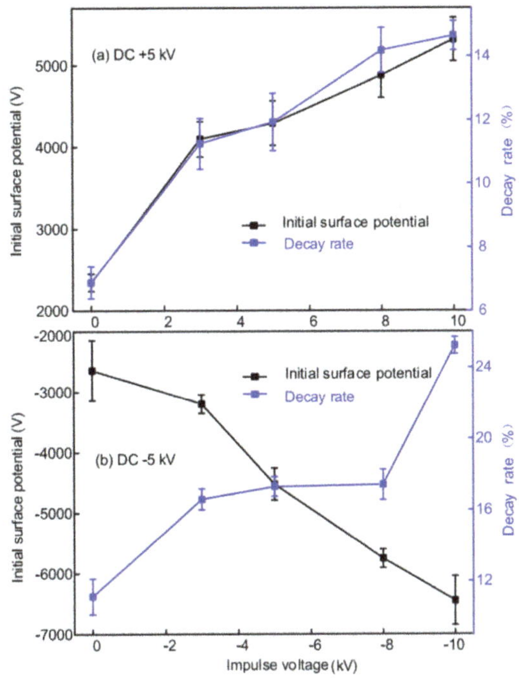

Figure 3. Charge injection processes under negative impulse voltage with a superimposed positive DC voltage

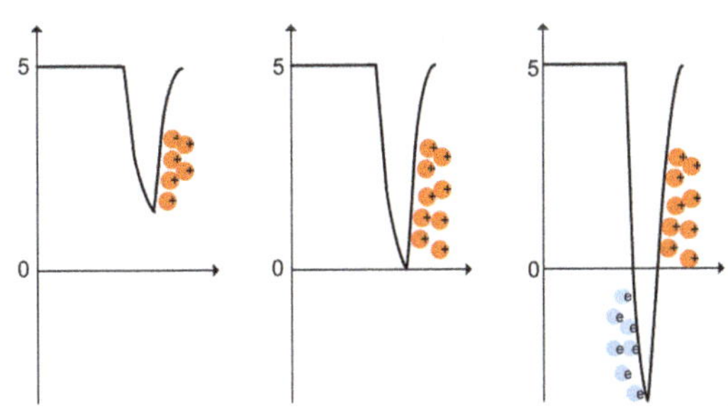

to some extent. When the pre-applied dc voltage is -5kV, the charge injection process is similar when the positive pulse voltage is superimposed. The decay rate of surface potential of the sample is related to the electric field inside the sample. The accumulation of surface charge forms the electric field inside the sample, so the decay rate of surface potential is synchronized with the initial change of surface potential.

Effect of Superimposed Voltage on Flashover

Flashover is related to carrier transport in dielectric between electrodes. During carrier transport, carriers trapped by shallow traps are easy to escape, while carriers trapped by deep traps are difficult to escape and tend to accumulate on the surface (Li, 2016). The accumulation of surface charge is an important factor to trigger flashover voltage. The superposition of pulse voltage increases the accumulation of surface charge, which has a great influence on the surface discharge characteristics of oil-paper insulation. In order to reveal the flashover mechanism of oil-paper insulation, flashover test was carried out on the sample under the compound electric field.

The experimental results are shown in the form of coordinate axes as shown in Figure 4. The abscissa indicates DC voltage and the ordinate indicates composite voltage. The first and third quadrants respectively represent the superposition of the positive DC voltage and the positive pulse voltage and the superposition of the negative DC voltage and the negative pulse voltage, and the second and fourth quadrants respectively represent the superposition of the negative DC voltage and the positive pulse voltage and the superposition of the positive DC voltage and the negative pulse voltage. In the first quadrant, the average value of the creeping discharge voltage of the single-added DC voltage is 18.8kV. When the pulse voltage is added, the creeping voltage can occur at 10, 12 or 14kV, that is, the superposition of the pulse voltage is greatly increased. The possibility of creeping discharge occurs. At the same time, it can be seen that as the DC voltage increases, the composite voltage gradually drops, which indicates that DC plays a key role in the occurrence of creeping discharge.

When the positive DC voltage is superimposed with the positive pulse voltage, the amount of surface charge increases, and more charge accumulates on the surface of the sample, and the surface is discharged from one electrode

Figure 4. Summary of flashover test results with the application of superimposed voltage

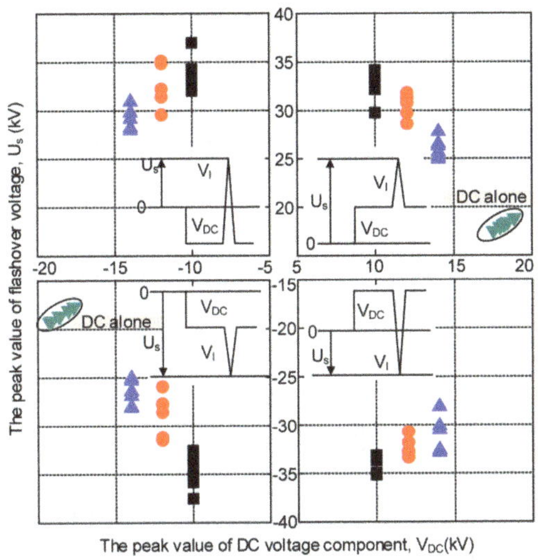

to the other through the trapping and detrapping process between the electrodes. In the third quadrant, the negative DC voltage and the negative pulse voltage are superimposed in the same manner as the first quadrant. In the second quadrant, the negative DC voltage is superimposed with the positive pulse voltage, and the composite voltage amplitude decreases as the DC voltage increases. And the composite voltage balance in the second quadrant is higher than the first quadrant because the negative charge injected by the pulse is combined with the pre-added positive charge, and the fourth quadrant is the same as the second quadrant. It can be seen from the four quadrants that the stimulation of the pulse voltage injects a large amount of electric charge between the two electrodes, which greatly improves the possibility of occurrence of oil-paper insulation creeping discharge.

EFFECT OF TEMPERATURE ON SURFACE CHARGE ACCUMULATION

Under DC Voltage

There are three ways to dissipate surface charges, including lateral diffusion, ion recombination, and bulk transport (Du, 2016). Since the thickness of the sample is thin, the vertical electric field excited by the surface charge is stronger than the horizontal electric field, so it is considered that the dissipating mode is mainly the third type.

The law of surface potential attenuation at different temperatures is shown in Figure 5. Figure 5 (a) corresponds to a DC voltage of 3.5kV, and Figure 5 (b) corresponds to a DC voltage of -3.5kV. It can be seen that the surface potential and the applied potential difference have the same polarity and decay with time, which indicates that the charge polarity of the polymerization on the surface of the sample has the same polarity as the applied field strength. Usually, there are two main potential dissipation stages, including the faster initial decay period and the after slower decay period at different temperatures (Du, 2013).

Figure 5 (a) shows the change of surface charge of oil-paper insulation with temperature under direct current voltage on positive polarity surface. It can be seen that when the temperature is 20 and 40°C, the curve of surface potential change with time is approximate to a straight line, but when 40°C, its slope is larger than 20°C, which indicates that the decay rate increases with the increase of temperature. When the temperature rises to 60°C, the surface potential decreases rapidly in a short time at the initial stage of attenuation, and then the attenuation curve gradually flattens out as time increases. When the temperature rose to 80°C, the surface potential experienced a initial rapid decay to zero. When negative direct current is applied and the temperature rises to 80°C, the positive potential attenuates to zero in the same time while the negative potential does not attenuate to the voltage. The attenuation trend at different temperatures is basically the same as that at the positive voltage. However, it can be seen that at the same temperature, the initial negative voltage indicates that the potential is higher than the initial point of the positive voltage. And when the temperature goes up to 80 °C, the positive potential falls to zero and the negative potential doesn't fall to zero in the same amount of time.

Figure 5. Relationship between the surface potential and decay time under different temperature

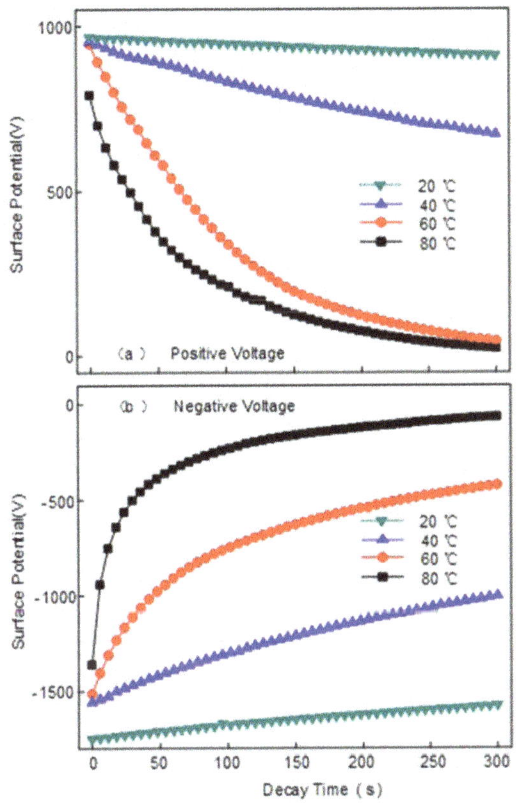

The potential decay rate of oil-paper insulation surface is shown in figure 6, which can better characterize the change of surface potential decay rate with the increase of temperature. Figure 6(a) shows the surface potential decay rate in figure 6(a). It can be seen that when the temperature is 20°C and 40°C, the decay rate decreases slightly with the increase of time but basically remains unchanged, except that the surface potential decay rate at 40°C is higher than that at 20°C. When the temperature increased from 20°C to 40°C, the decay rate increased at all times, indicating that the increase in temperature accelerated the decay of surface potential and promoted the dissipation of charge. Similarly, when the temperature rises to 60°C and

Figure 6. Relationship between the surface potential decay rate and decay time under different temperature

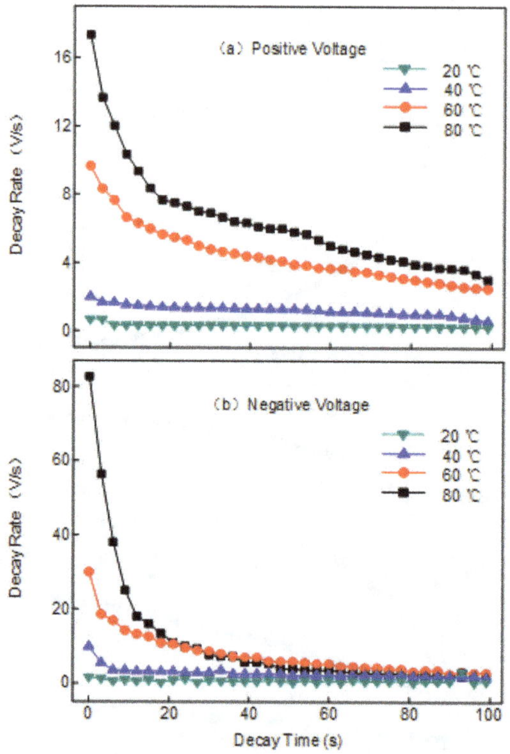

80°C, the attenuation rate increases obviously, especially in the initial stage of attenuation, when the attenuation rate reaches the maximum. At 60°C and 80°C, the change of decay rate with time can be divided into two stages, namely, the rapid decay stage at the beginning of decay and the subsequent decay rate basically unchanged. This is shown in figure 5: at the initial stage of attenuation, the surface potential rapidly decays, and with the increase of time, the surface potential decays slowly. In figure 6 (b), the change trend of surface potential decay rate when negative voltage is applied is basically the same as that when positive voltage is applied, but it can be seen that in the early stage of attenuation, the decay rate under negative voltage is faster than that under positive voltage.

Under Superimposed Voltage

The change of ambient temperature greatly affects the charge distribution characteristics of dielectric surface. In this chapter, the +5kV DC voltage is applied for 3min, and the ±10kV pulse voltage is applied for 10s. The temperature of the control station is 20°, 40°, 60° and 80°C. The law of surface potential attenuation at different temperatures is shown in Figure 7. The surface charge dissipation at different composite voltages has the same trend with temperature, which is not listed in this chapter. Note that under the action of temperature, when the sample is subjected to the composite voltage of DC or DC voltage and pulse voltage, the surface potential attenuation trend is basically the same. Obviously, when the same polarity voltage is superimposed, the surface potential is higher than the surface potential when

Figure 7. Relationship between the surface potential and the decay time under ±10 kV impulse with a superimposed 5 kV DC voltage at the temperature of 20, 40, 60 and 80 °C

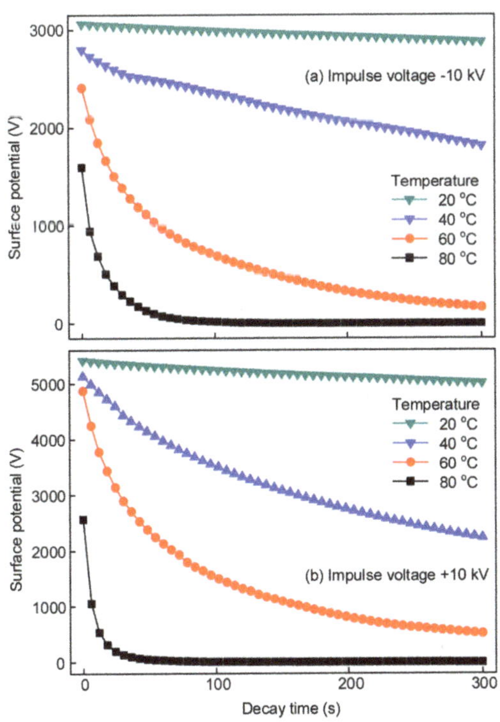

the opposite polarity voltage is superposed. The superimposed pulse voltage in figure 7(a) is -10kV. When the heating stage temperature is 20°C, the amount of surface potential change during recording time is small. As the temperature increases, the amount of surface potential changes increases, and when the temperature rises by 80°C, the surface potential quickly decays to zero within five minutes. In addition, the initial surface potential drops as the temperature increases. In figure 7(b), the superimposed pulse voltage is +10kV, and the surface potential changes with temperature is basically the same as in figure 7(a). This shows that under the superimposed electric field, temperature still plays a major role in surface charge accumulation and dissipation. The increase in temperature reduces the accumulation of surface charge and accelerates the dissipation of surface charge.

EFFECT OF TEMPERATURE ON TRAP DISTRIBUTION

Under DC Voltage

The accumulation and dissipation process of surface charge is closely related to the charge trapping-detrapping process, so it is important to study the trap distribution characteristics of oil-paper insulation samples (Du, 2017). According to the measured surface potential attenuation data and the calculation method of dielectric trap distribution, the surface voids and electron trap distribution of the sample at different temperatures can be obtained. Typical trap distributions have double peaks, representing shallow traps and deep traps, respectively. The accumulation of charge trapped on the surface of the sample caused a change in surface potential. The charge trapped by the shallow trap is more likely to be trapped more easily, and the charge trapped by the deep trap takes a long time to fall out. This is also the reason why the surface potential is initially decayed faster and the decay rate is gradually slowed down (Du, 2018).

Figure 8 and 9 respectively represent the trap energy level and trap density of shallow traps of electrons and holes at different temperatures. As can be seen from figure 8, with the increase of temperature, the energy level of the shallow hollow trap increases from ~0.81 eV to ~0.88 eV, and the energy level of the shallow hollow trap increases from ~0.76 eV to ~0.84 eV. The energy level of the shallow hollow trap is higher than that of the shallow

Figure 8. Relationship between the shallow trap level and temperature under positive and negative voltage

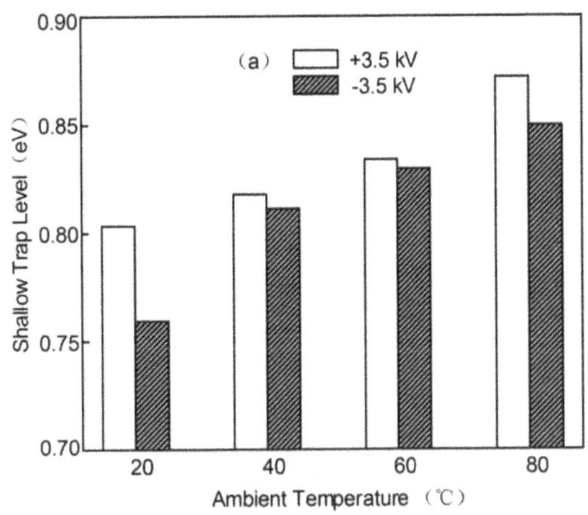

Figure 9. Relationship between the shallow trap density and temperature under positive and negative voltage

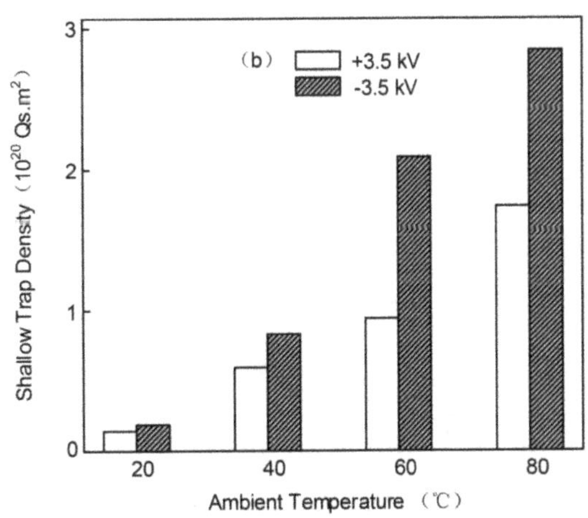

hollow trap. In figure 9, the density of both hole and electron shallow traps increases with the increase of temperature, and the density of electron shallow traps is higher than that of hole shallow traps.

The columnar diagram of the deep trap energy level and density of electrons and holes varying with temperature is shown in figure 10 and 11. Similarly, the energy levels of electrons and deep traps increase with the increase of temperature. The electron deep trap energy level increased from ~0.87eV

Figure 10. Relationship between the deep trap level and temperature under positive and negative voltage

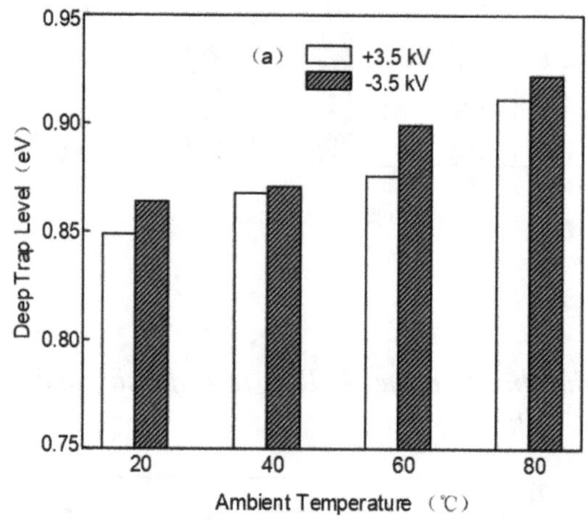

Figure 11. Relationship between the deep trap density and temperature under positive and negative voltage

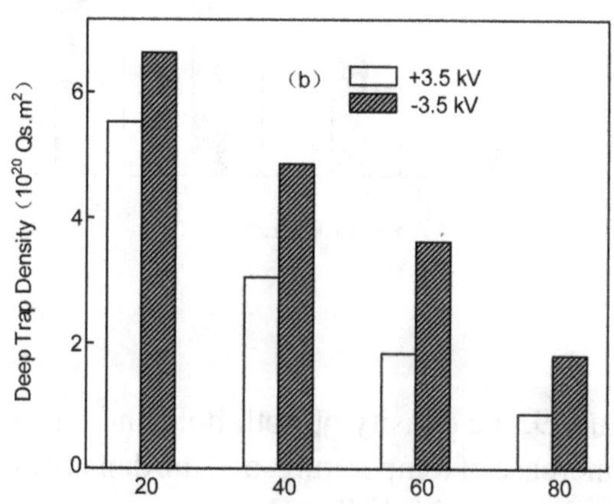

to ~0.92eV, and the hole deep trap energy level increased from ~0.85eV to ~0.91eV. The electron deep trap energy level was higher than the hole deep trap energy level. However, it can be seen from figure 11 that the density of electron and hole deep traps decreases with the increase of temperature, and the density of electron deep traps is higher than that of hole traps.

As the temperature increases, the electron or hole will gain more energy. Some charge energy is higher than the trap barrier and will not be captured by the trap. Some charge can easily escape from the trap due to its high kinetic energy. In figure 8 and 10, the energy levels of both holes and electron shallow traps and deep traps increase with the increase of temperature. This is because holes or electrons with higher kinetic energy are more likely to escape from the traps, and some electric charges that are difficult to escape at low temperature can escape at high temperature, so they are considered as shallow traps at high temperature. For example, the hole deep trap level is -0.85eV in figure 10 at 20°C, while its shallow trap level is -0.88eV at 80°C in figure 8. In other words, the hole caught in the deep hole trap at low temperature can easily escape at high temperature and is considered to be caught by the shallow hole trap. When deep traps are converted into shallow traps, the number of shallow traps increases and the density of shallow traps increases, as shown in figure 9. While there are charges in deeper traps that can escape, the number is small and many deep traps are converted to shallow traps, so the number of deep traps is reduced, as shown in figure 11.

Figure 10 and 11 reveal the variation of the energy of deep well and the density of homologous well with the temperature. Similarly, charges trapped in deeper traps with higher energy levels are more likely to escape at higher temperatures. Thus the traps with higher energy levels can be measured, represented by the height of the deep trap energy level as shown in figure 10. However, under positive and negative dc voltages, the density of the deep well decreases as the temperature increases. Although deeper trap levels have been measured, the number is limited, and many deep traps have been redefined as shallow traps. As a result, less charge is captured in the deep trap and less charge can be measured to escape from the deep trap.

The shallow trap density of holes and electrons increases and the density of deep traps decreases, indicating that more charges are trapped by shallow traps and are more likely to collapse. Therefore, the surface potential decay rate increases with temperature. The trap density of electrons at the same temperature is higher than that of the hole trap because electrons are more likely to be generated during corona discharge, and more electrons are trapped by traps. Therefore, the electron trap density is higher than the hole trap density. The initial surface potential at voltage is higher than the positive voltage.

Carrier mobility is closely related to the traping-detraping process and dissipation of surface charge. Figure 12 describes the relation of electron and hole carrier mobility with temperature at different temperatures. With the increase of temperature, the mobility of hole and electron carrier increases, and the mobility of hole is higher than that of electron. The change of carrier mobility clearly indicates that the carrier gains more energy when the temperature increases, and it is easier for the carrier to migrate from the surface to the ground electrode, thus the surface charge dissipation rate is faster. Under the same voltage amplitude, the charge accumulation under negative voltage is more serious and difficult to dissipate.

Under Superimposed Voltage

The accumulation and dissipation process of surface charge is closely related to the charge trapping-detrapping process, so it is important to study the trap distribution characteristics of oil-paper insulation samples. Combined with the surface potential

measurement results and the dielectric trap calculation method, the distribution of sample holes and electron traps at different temperatures can

Figure 12. Relationship between the carrier mobility and temperature under positive and negative voltage

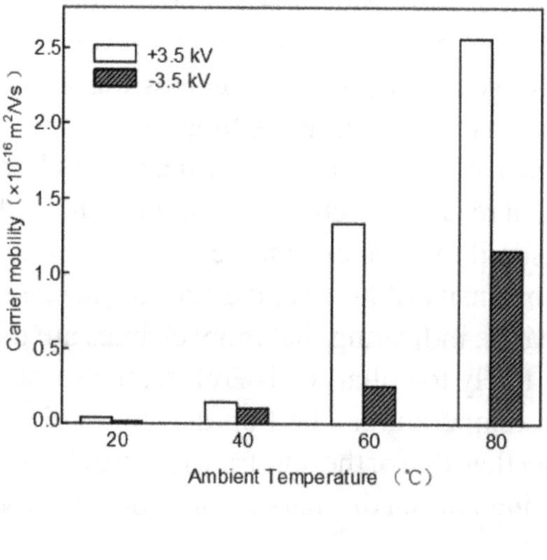

be obtained. The accumulation of charge trapped on the surface of the sample caused a change in surface potential. The relationship between trap density and temperature is shown in Figure 13. Under positive and negative pulse voltages, as the temperature increases, the shallow trap density increases and the deep trap density decreases (Du, 2019). The density of deep traps decreases with increasing temperature, and the density of deep traps under the same polarity superimposed voltage is greater than the density of hole traps under different polarities. At the same time, the deep trap at low temperature is transformed into a deep trap at high temperature, so the shallow trap density increases and the deep trap density decreases.

Figure 13. Relationship between the trap density and the temperature under ±10 kV impulse voltage with a superimposed 5 kV DC voltage

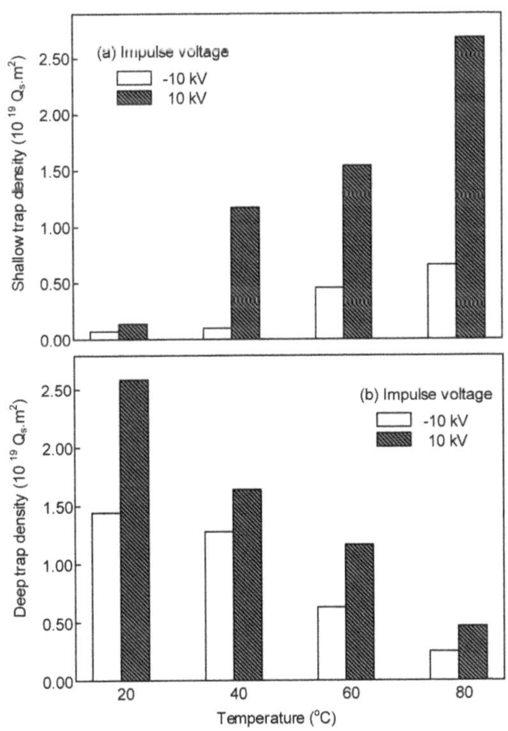

EFFECT OF TEMPERATURE ON TRAP DISTRIBUTION

Under DC Voltage

The surface trap characteristics have an important influence on the surface discharge performance of insulators. In this chapter, the double-finger electrodes are used to study the surface discharge characteristics of oil-paper insulation at different temperatures under DC voltage. In the experiment, the oil-paper insulation samples were subjected to positive and negative pressure respectively, and the measured creeping discharge voltage results were calculated by Weibull probability distribution. The experimental results are shown in Figure 14. The results show that as the temperature increases,

Figure 14. Weibull plots of DC surface discharge strength of the samples with different temperature under positive and negative voltage

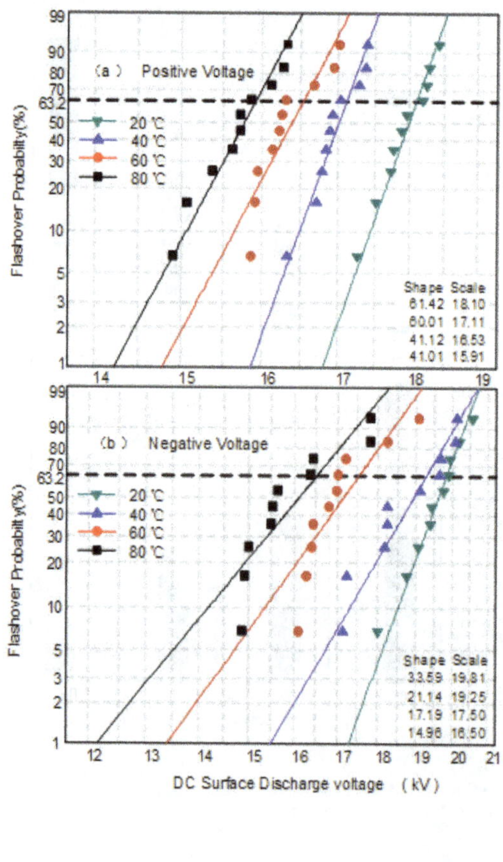

the creeping discharge voltage of the oil-paper insulation under both positive and negative voltages decreases. It can be seen that the creeping discharge voltage of the oil-paper insulation under positive voltage is lower than the creeping discharge voltage under negative voltage, that is to say, the oil-paper insulation is more prone to creeping discharge under positive DC voltage. According to the above research on the trap distribution characteristics at different temperatures, as the temperature increases, the trap levels of deep traps and shallow traps increase, the shallow trap density increases and the deep trap density decreases (Du, 2019). Note that the charge is more easily the shallow trap captures and escapes from the trap, triggering a secondary electron avalanche of the oil-paper insulation surface, resulting in a reduced creeping discharge voltage.

As can be seen from Figures 12, the increase in temperature causes the carrier mobility to increase, which means that the charge has sufficient kinetic energy to reach the ground electrode from the surface of the sample, so the kinetic energy of the charge increases at high temperatures, so that In the case of creeping discharge, from one electrode to the other, in this case, creeping discharge is more likely to occur. In addition, the creeping discharge voltage under positive DC voltage is lower than that under negative DC voltage. This is because there is a difference between the electron trap and the hole trap barrier. As shown in Figure 10, the electron deep trap level is required. It is higher than the hole deep trap level so that the electrons are less likely to collapse.

Under Superimposed Voltage

Figure 15 shows the Weibull distribution of the superimposed flashover voltage at different temperatures at 10kV DC. Figure 15(a) corresponds to the creeping discharge voltage at the same polarity combined voltage, and Figure 15(b) corresponds to the creeping discharge voltage at the heteropolar composite voltage. The results show that with the increase of temperature, the creeping discharge voltage of the oil-paper insulation under the same polarity composite voltage and the heteropolar composite voltage decreases. According to Figures 13(a) and 13(b), as the temperature increases, the shallow trap density increases and the deep trap density decreases. As the temperature increases, the number of shallow traps for oil-paper insulation

Figure 15. Weibull plots of the flashover voltage of samples at different temperatures under impulse voltage with a superimposed 10 kV DC voltage

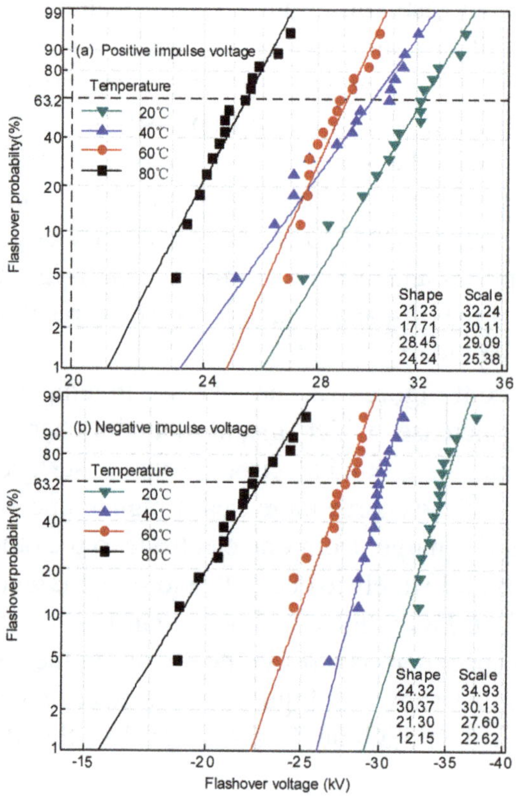

increases, and more charge can escape from the trap to participate in the creeping discharge process, making the surface discharge more likely to occur at a higher temperature.

CONCLUSION

This chapter studies the surface charge and flashover characteristics of oil-paper insulation used in converter transformers. The surface charge and flashover behavior of oil-paper insulation samples under electrothermal composite field were studied, and it was found that the introduction of pulse voltage would increase the probability of surface discharge, and the surface discharge would occur due to overvoltage when the converter transformer was running at a small DC voltage. At the composite voltage, the effect of temperature on the surface charge and creeping discharge of the oil-paper insulation is substantially the same as under DC voltage. No matter the polarity of dc voltage is positive or negative, the initial surface potential decreases with the increase of temperature, the surface charge dissipates faster, and the surface discharge voltage decreases. Only at a higher temperature, when the heteropolar DC voltage and the pulse voltage are combined, the turning point when the initial surface potential starts to decrease with the increase of the pulse voltage is earlier, which indicates that in the electrothermal compound field, temperature has a more obvious influence on the motion characteristics of surface charge.

REFERENCES

Chao, T., Chen, G., Fu, M., & Liao, R. J. (2010). Space charge behavior in multi-layer oil-paper insulation under different dc voltages and temperatures. *IEEE Transactions on Dielectrics and Electrical Insulation*, *17*(3), 775–784. doi:10.1109/TDEI.2010.5492250

CIGRE. (2010). *HVDC converter transformers-design review, test procedures, ageing evaluation and reliability in service*. JWG A2/B4.28.

Du, B., Chang, R., Zhu, W., Li, J., & Jiang, J. (2019). Temperature-dependent surface charge and discharge behaviour of converter transformer oil–paper insulation under dc voltage. *IET Science, Measurement & Technology*, *13*(1), 29–34. doi:10.1049/iet-smt.2018.5018

Du, B. X., Chang, R., Jiang, J., & Li, J. (2018). Temperature-dependent surface charge and flashover behaviors of oil-paper insulation under impulse with superimposed dc voltage. *IEEE Access: Practical Innovations, Open Solutions*, 1–1.

Du, B. X., Du, Q., Li, J., & Liang, H. C. (2018). Carrier Mobility and Trap Distribution Dependent Flashover Characteristics of Epoxy Resin. *IET Generation, Transmission & Distribution*, *12*(2), 466–471. doi:10.1049/iet-gtd.2017.0984

Du, B. X., Jiang, J. P., Zhang, J. G., & Liu, D. S. (2016). Dynamic behavior of surface charge on double-layer oil-paper insulation under pulse voltage. *IEEE Transactions on Dielectrics and Electrical Insulation*, *23*(5), 2712–2719. doi:10.1109/TDEI.2016.7736830

Du, B. X., & Li, J. (2016). Effects of ambient temperature on surface charge and flashover of heat-shrinkable polymer under polarity reversal voltage. *IEEE Transactions on Dielectrics and Electrical Insulation*, *23*(2), 1190–1197. doi:10.1109/TDEI.2015.005182

Du, B. X., & Li, J. (2017). Surface charge coupling behavior of fluorinated polyimide film under dc and pulse voltage. *IEEE Transactions on Dielectrics and Electrical Insulation*, *24*(1), 567–573. doi:10.1109/TDEI.2016.005988

Du, B. X., Li, J., & Du, W. (2013). Surface charge accumulation and decay on direct-fluorinated polyimide/Al_2O_3 nanocomposites. *IEEE Transactions on Dielectrics and Electrical Insulation*, *20*(5), 1764–1771. doi:10.1109/TDEI.2013.6633707

Du, B. X., Li, X. L., & Jiang, J. P. (2016). Surface charge accumulation and decay on directfluorinated oil-impregnated paper. *IEEE Transactions on Dielectrics and Electrical Insulation*, *23*(5), 3094–3101. doi:10.1109/TDEI.2016.7736874

Du, B. X., Xu, R. R., Jin, L., & Li, Z. L. (2018). Improved carrier mobility dependent surface charge am flashover voltage of polypropylene film under dc and pulse voltages. *IEEE Transactions on Dielectrics and Electrical Insulation*, *25*(3), 1014–1021. doi:10.1109/TDEI.2018.006845

Du, B. X., & Zhang, J. G. (2016). Charge coupling behavior of double-layer oil-paper insulation under dc and pulse voltages. *IEEE Transactions on Dielectrics and Electrical Insulation*, *23*(4), 1–9. doi:10.1109/TDEI.2016.7556479

Du, B. X., Zhu, W. B., Li, J., Xing, Y. Q., & Huang, P. H. (2017). Temperature-dependent surface charge behavior of polypropylene film under dc and pulse voltages. *IEEE Transactions on Dielectrics and Electrical Insulation*, *24*(2), 774–783. doi:10.1109/TDEI.2017.006290

Flourentzou, N., Agelidis, V. G., & Demetriades, G. D. (2009). Vsc-based hvdc power transmission systems: An overview. *IEEE Transactions on Power Electronics*, *24*(3), 592–602. doi:10.1109/TPEL.2008.2008441

Jadidian, J., Zahn, M., Lavesson, N., Widlund, O., & Borg, K. (2012). Surface flashover development on transformer oil-pressboard interface. In *Power Modulator and High Voltage Conference (IPMHVC), 2012 IEEE International*. IEEE. 10.1109/IPMHVC.2012.6518674

Leblanc, P., Paillat, T., Morin, G., & Perrier, C. (2015). Behavior of the charge accumulation at the pressboard/oil interface under dc external electric field stress. *IEEE Transactions on Dielectrics and Electrical Insulation*, *22*(5), 2537–2542. doi:10.1109/TDEI.2015.005108

Li, G., Chen, G., & Li, S. (2016). Charge Transport Characteristics in Nanodielectric. *IEEE Conference on Electrical Insulation & Dielectric Phenomena*.

Okabe, S., Ueta, G., Utsumi, T., & Nukaga, J. (2015). Insulation characteristics of gis insulators under lightning impulse with dc voltage superimposed. *IEEE Transactions on Dielectrics and Electrical Insulation*, *22*(6), 1–9. doi:10.1109/TDEI.2015.005178

Sima, W., Sun, P., Yang, M., Yang, Q., & Wu, J. (2015). Effect of space charge on the accumulative characteristics of oil paper insulation under repeated lightning impulses. *IEEE Transactions on Dielectrics and Electrical Insulation*, *22*(5), 2483–2490. doi:10.1109/TDEI.2015.004843

Vasa, A. N. N. J., Vinu, R., & Sarathi, R. (2017). Influence of ambient medium on thermal ageing of pressboard in transformer oil containing dibenzyl bisulphide (dbds). *IEEE Transactions on Dielectrics and Electrical Insulation*, *24*(5), 2836–2846. doi:10.1109/TDEI.2017.006412

Wang, S. Q., Zhang, G. J., Mu, H. B., Wang, D., Lei, M., Suwarno, S., & … . (2012). Effects of paper-aged state on space charge characteristics in oil-impregnated paper insulation. *IEEE Transactions on Dielectrics and Electrical Insulation*, *19*(6), 1871–1878. doi:10.1109/TDEI.2012.6396942

Zhou, H. Y., Ma, G. M., Li, C. R., Shi, C., & Qin, S. C. (2017). Impact of temperature on surface charges accumulation on insulators in sf6-filled dc-gil. *IEEE Transactions on Dielectrics and Electrical Insulation*, *24*(1), 601–610. doi:10.1109/TDEI.2016.005838

Zhou, Y., Jin, F., Huang, M., Sha, Y., & Huang, J. (2013). Influence of temperature on developing process of surface flashover in oil-paper insulation under combined AC-DC voltage. In *2013 IEEE Conference on Electrical Insulation and Dielectric Phenomena - (CEIDP 2013)*. IEEE. 10.1109/CEIDP.2013.6748251

Chapter 5
Trap Property and Charge Transmission in PE

ABSTRACT

The electrical properties of the dielectric are achieved by affecting the charge transfer process. The trap characteristics have an important influence on the electrical properties of the dielectric by affecting the charge transfer process. Aggregation and trap level characteristics of nanographene on low density polyethylene (LDPE). The direct current conductivity, breakdown strength, trap level distribution, space charge distribution, and charge mobility of nanocomposites were investigated. The experimental results show that the interface region between graphene and polymer introduces many deep traps in the forbidden band of nanocomposites, which can reduce the trapping process of charge and inhibit the accumulation of space charge. This indicates that the addition of nanoscale graphene has a significant improvement in the electrical performance of high voltage DC cables, which will provide a reference for production and application.

INTRODUCTION

In 1994, the concept of nano-dielectrics was proposed, which has become a research hotspot in the field of electrical and electronic engineering (Lewis, 1994). Numerous studies have shown that nanocomposites have better electrical, thermal and mechanical properties than the original polymers. Nanocomposites (Nelson, 2002; Tanaka, 2004; Nelson, 2005; Krivda, 2012;

DOI: 10.4018/978-1-5225-8885-6.ch005

Tanaka, 2013), with excellent electrical properties, have low electrical conductivity, high breakdown strength and enhanced space charge resistance, making it These new properties as a new generation of insulating materials are believed to be due to the high specific surface area of the nanofillers resulting in the formation of large amounts of deep traps in the polymer-filler interface region (Nelson, 2004; Lewis, 2004; Tanaka, 2005).

In previous studies, including metal oxides (such as alumina, silica, titania, magnesia, and zinc oxide), nitrides (such as boron nitride and aluminum nitride), montmorillonite (MMT), etc. The filler is incorporated into the polyolefin (Kumara, 2016; Tanaka, 2011; Zha, 2008; Jung, 2010; Li, 2014). In addition, the surface modification of the nanofiller can reduce the agglomeration of the nanofiller and achieve better dispersibility in the polymer matrix. Therefore, as a unique nanoscale filler, graphene has only the thickness of the atomic layer and has up to 2630. The huge specific surface area of m2 / g enhances the interfacial region of the polymer filler (Huang, 2009; Lau, 2013; Li, 2013; Rafiee, 2009; Li, 2013; Fim, 2013), which may significantly increase the interfacial area of the polymer-filler and exert greater insulation potential.

For nanocomposites using graphene, previous research has focused on thermal, mechanical, and electrical properties. Graphene/LDPE nanocomposites are reported to be semiconductors when the filler content exceeds a critical percolation threshold of 3.8% by volume (Fim, 2013). 1 wt% Graphene/LDPE nanocomposites have enhanced mechanical properties and non-linear conductivity (Gaska, 2017). So far, little research has been done on the electrical properties and trap level characteristics of graphene-doped nanocomposites with extremely low filler content. According to the operating temperature of polyethylene insulation in HVDC transmission cables, it is necessary to study the effects of ambient temperature on charge transfer and trap level characteristics.

In terms of electrical properties, space charge is usually caused by the charge injected by the electrode and the ionization of chemical groups up to tens of kilovolts/mm under a DC electric field (Wu, 2017; Montanari, 2011; Li, 2016). This will represent the electric field distribution and accelerate the aging, degradation or even destruction of the insulation (Lan, 2014; Wang, 2016). The occurrence of voltage levels, the miniaturization of electrical equipment and the increase in operating temperature will further increase the space charge injecting and accumulating, which has become one of the key issues in the development of pe-based insulation.

CHARGE CARRIER TRANSPORT MODULATED BY TRAP

In this study, LDPE was used as the polymer matrix. LDPE/graphene NCs with the filler content of 0, 0.005, 0.01, 0.1 and 0.5 wt% are prepared. The breakdown strength, conductivity, space charge behavior, and trap level distribution of the NC were measured. According these results, a schematic model is proposed to illustrate that charge carrier transport process modulated by trap, which further reveals the relationship between electrical properties and trap energy level distribution.

Characterization of NCs

Figures 1 a-d are SEM images of 0, 0.005, 0.01 and 0.1 wt% of LDPE / graphene NC. It can be seen that when the filler content is 0.005 and 0.01% by weight, the graphene filler is sufficiently dispersed in the polymer matrix. As the volume fraction of the graphene filler continues to increase, the closest distance (NDNF) between adjacent fillers gradually decreases. The SEM image of 0.1 wt% NCs shows that NDNF became submicron.

Figure 1. SEM images of 0, 0.005, 0.01, and 0.1 wt % LDPE/graphene NC cross sections correspond to (a, b, c, and d), respectively. (e) The relationship between the DSC curve of LDPE/graphene NCs and the filler content.

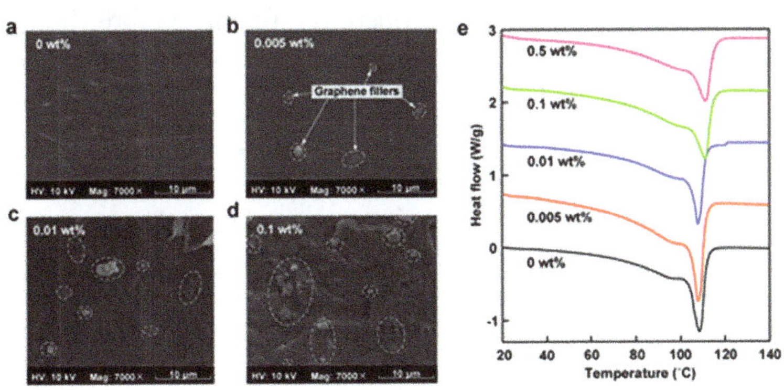

Figure 1e is a DSC curve for LDPE / graphene NC. The crystallinity (X) can be obtained by integrating the DSC curve in the following manner (Gray, 1970):

$$X=\frac{\Delta H_m}{\Delta H_0}\times 100\% \tag{2-1}$$

Where ΔH_0 is the melting enthalpy (100%) of the fully crystalline LDPE, usually ΔH0 = 293 J / g (Peng, 2016). ΔH_m is the melting enthalpy of the LDPE NC studied. The melting enthalpy and temperature are shown in Table 1. The crystallinity of the samples of 0, 0.005, 0.01 and 0.1 wt% as determined by DSC were 32.6%, 39.5%, 37.0%, 34.6% and 32.5%, respectively. When the graphene content is 0.005 wt%, the material crystallinity is the highest because of heterogeneous nucleation. The reason is that a well-dispersed graphene filler become a heterogeneous nucleating agent which improves the crystallinity of the LDPE composite. However, when the filler content reaches 0.01 wt%, the graphene filler restricts the movement of the polymer molecular chain, and there is almost no excessive crystallization space, resulting in a decrease in crystallinity. The results show that when the filler content exceeds 0.1%, the melting point temperature rises slightly.

Breakdown Strength and Conductivity

Figure 2a shows the relationship between the conductivity of LDPE/graphene with a polarization field of 10 kV / mm. Especially for samples of 0.005 wt% and 0.01 wt%, the charging current dropped significantly before 100th seconds. After 100s, the charging current reaches a steady state and the decay rate becomes slow. The reason is that the charging current include the conduction

Table 1. Melting enthalpy, temperature and crystallinity levels

Specimens	Melting Temperature (°C)	Crystallinity (%)	Melting Enthalpy ($J\cdot g^{-1}$)
0 wt%	106.1	32.6	95.63
0.005 wt %	105.6	39.5	115.72
0.01 wt%	105.4	37.0	108.35
0.1 wt%	108.6	34.6	101.41
0.5 wt%	108.9	32.5	95.15

Figure 2. (a) DC conductivity and (b) DC breakdown strength

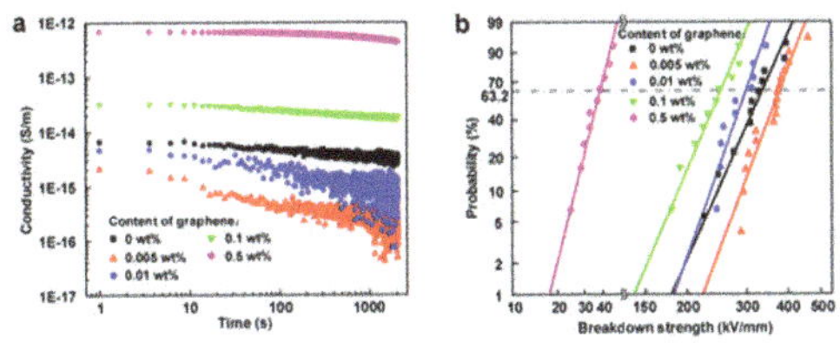

and polarization current. As the polarization time increases, the polarization gradually enters the steady state, and the conduction current plays a major role. However, the charge current in 0.5 wt% composites will slowly decay during the measurement time until it reaches a steady state of 2000 s due to the injection and accumulation of space charge.

The study represents the DC conductivity with an average conductivity of the last 100 seconds, 3.19×10^{-15} S / m, 1.99×10^{-16} S / m, 7.58×10^{-16} S / m, 2.03×10^{-14} S / m and 4.88×10^{-13} S / m respectively, corresponding to the weight percentage of the LDPE / graphene composite at 0, 0.005, 0.01 and 0.1 wt% 0.5 wt%. Graphene has high electrical conductivity under electrical stress. As shown in Fig. 1b, the filler of 0.005 wt% is completely dispersed. It is shown that the 0.005 wt% filler is much lower than the permeability threshold, so that no conductive path is formed even in a high electric field. In addition, graphene fillers have a large specific surface area and a large interaction area with the LDPE matrix, which traps carriers and inhibits their transport. Adding these microscopic effects together causes DC conductivity decreasing.

Figure 2a also shows that the DC conductivity increases as the filler content increases from 0.005 wt% to 0.1 wt%. As shown in Figures 1a to d, NDNF gradually decreases as the filler content increases. When the content is 0.1% by weight, the NDNF reaches the submicron or nano level, and therefore, the polymer-filler interaction regions can overlap each other, thereby providing a low resistance path for electrons and holes and accelerating the local transport of carriers. This leads to an increase in conductivity. When the content is further increased to 0.5 wt%, the filler content is above the percolation threshold, resulting in a thermally activated field at the polymer-filler interface that enhances the carrier transition under applied stress. As shown in Figure 2a, the thin polymer layer between adjacent fillers will be passed through by carriers, which resulting in a sharp increase in DC conductivity.

Figure 2b shows the Weibull distribution of the DC breakdown strength. The breakdown strengths at 0, 0.005, 0.01, 0.1 and 0.5 wt% are 336.4, 379.5, 301.4, 252.8 and 38.7 kV / mm. Samples with lower DC conductivity were found to have larger DC breakdown strength. The maximum breakdown strength of 0.005 wt% of the composite is due to the inhibition of charge carrier transport by the polymer-filler interaction zone. As the filler content is further increased from 0.005% by weight to 0.5% by weight, transport of carriers through the amorphous regions is facilitated, which involves capture and decapture processes. Through the non-radiative transition of energy in the Auger-type process, the trapping and recombination of charge carriers will generate hot electrons that collide with the molecules and dissociate into free radicals. The process can be expressed as:

$$\begin{aligned} AB + e^-(\text{hot}) &\longrightarrow A^* + B^* + e^-(\text{cold}) \\ \text{or} &\longrightarrow A^* + B^* + e^-(\text{trapped}) + \text{energy release} \end{aligned} \tag{2-2}$$

This process will promote the formation of low density regions, making electrons more susceptible to acceleration by the electric field and kinetic energy to cause molecular chain rupture, which in turn causes partial discharge and further decomposition.

Space Charge

The space charge distribution and electric field distribution (polarized electric field 50 kV/mm) of the LDPE/graphene composite with a content of 0, 0.005, 0.01 and 0.1 wt% in 1800 seconds and the space charge of pure LDPE are showed in Figure 3. As the polarization process progresses, both charge peaks close to the electrode move toward the polymer, indicating that both electrodes are implanted with a large amount of the same polarity charge. The interface between the crystalline phase and the amorphous phase of the LDPE introduces a trap, thereby trapping the charge and causing the accumulation of space charge. The results show that in 1800 seconds, a large amount of electrons is injected to a deep depth and captured throughout the LDPE, since the mobility of electrons is much higher than the mobility of holes.

Comparing the space charge in Figures 3a and b, it was found that space charge accumulation was less in the 0.005 wt% sample and the electric field was

Figure 3. Space charge distribution (polarized electric field 50 kV/mm) and electric field distribution. The contents are (a) Neat (b) 0.005 wt%. (c) 0.01 wt%. and (d) 0.1 wt%.

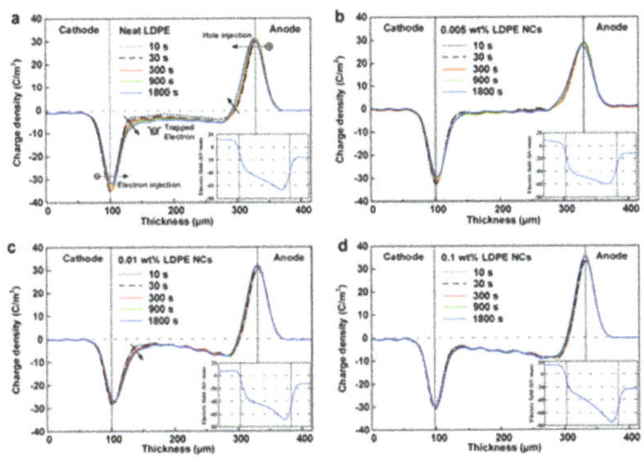

more uniform. The polymer-graphene interaction zone inhibits the transport of charge carriers, resulting in a large amount of co-polar charge clustering near the two electrodes. The effective electric field at the electrode-polymer interface is reduced, the barrier of the Schottky implant is enhanced, and the injection of the same polarity charge from the electrode into the polymer is inhibited.

However, a further increase in the filler content will result in an increase in space charge accumulation. Figures 3c and shows that severe electric field distortion occurs near the anode. A large amount of electrons is injected into the polymer body because the overlapping positions between the molecular chains will provide a resistance path for electrons, thereby accelerating the transport through the chain barrier and causing electron migration.

Trap

The above experimental results show that 0.005 wt% NCs have less inhibition of carrier transport than pure LDPE, lower DC conductivity, higher breakdown strength and less space charge accumulation. The IDC test was used to characterize the trap level distribution. he relationship between the isothermal discharge current (I) of NCs and the depolarization time (t) are showed in Figure 4(a). It is found that as the depolarization process progresses, the discharge current gradually decreases, which is caused by the deepening of traps involved in the charge release mechanism. According to the IDC theory, the traps density ($N_t(E)$) and traps energy (E_t) can be calculated as:

$$N_t(E) = \frac{2dIt}{el^2kT} \tag{2-3}$$

$$E_t = kT\ln(\nu t) \tag{2-4}$$

Where T is the absolute temperature, d is the thickness of the film, e is the electron charge, k is the Boltzmann constant, l is the penetration depth of the injected electrons, and ν is the escape frequency of the trapped electrons, equal to 10^{12} s^{-1}in the LDPE.

The trap level distribution of LDPE/graphene NC is shown in Figure 4b. A deep trap of all LDPE/graphene NC was observed to be located at approximately 0.95 eV. Studies have shown that adding alumina-doped epoxy resin (EP / Al^2O^3), silica-doped silicone rubber (SiR / SiO^2) and titanium oxide-doped polyimide (PI / TiO^2) A deep charge trap will be introduced. Tanaka proposed a typical multi-core model of spherical nanoparticles, in which the polymer-graphene interface consists of a bonding layer, an adhesive layer, and a loose

Figure 4. (a) The relationship between the isothermal discharge current and depolarization time. (b) Trap level distribution (c) Deep modulated charge carrier transport (d) The transport of charge carriers through the overlap of the interaction zones and the thermal activation field between adjacent fillers enhances the transition.

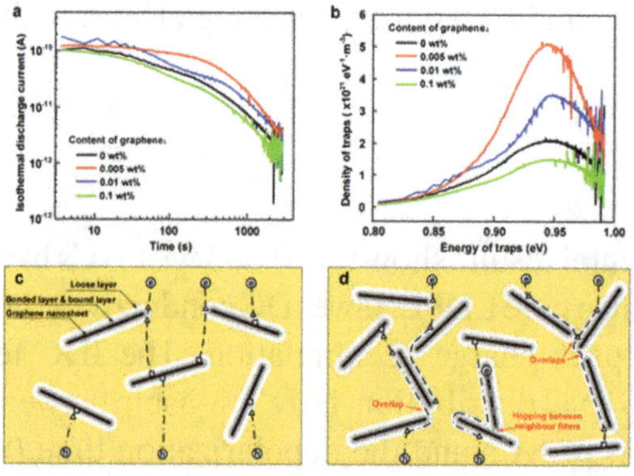

layer. A large number of deep traps of electrons and holes are distributed in the bonding and bonding layers, which trap the charge in the polymer and increase the average jump barrier of the charge carriers, thereby suppressing the transport of charge carriers. The density of the deep trap increases first and then decreases as the filler content increases. As shown in Figures 4c and d, with reference to the multi-core model, a plate-like interaction zone appears at the interface between the filler and the matrix.At the same mass fraction, graphene fillers have a much larger specific surface area than the traps provided by other nanoparticles. Figure 3b shows that deep trap-modulated carrier trapping results in the accumulation of the same polarity charge near the electrode, thereby enhancing the barrier of Schottky injection, increase the electric field for chargeinjection and suppress the accumulation of space charge. In addition, the reduction of the injected charge can suppress charge trapping or recombination, reduce the formation of high-energy electrons and improve the breakdown strength.

As the filler content is further increased to 0.1 wt%, the NDNF becomes submicron or nanoscale, as shown in Figure 1d. The loose layers of the outermost layers of the nanoparticles will overlap each other, allowing electrons and holes to pass through this low resistance path, as shown in Figures 4d. Furthermore, at locations where the polymer layer between adjacent fillers is sufficiently thin, typically within a few tens of nanometers, a large number of charge carriers can be jumped by thermally activated field-enhancing hops between adjacent fillers. The thin polymer layer accelerates the local transport of carriers, further leading to an increase in DC conductivity. The accumulation of space charge can also improve the energy of hot electrons, accelerate the formation and even growth of micropores in the amorphous region, resulting in a decrease in breakdown strength.

GRAPHENE INHIBIT THE SPACE CHARGE INJECTION AND ACCUMULATION

Polyolefin insulated dielectrics tend to accumulate space charge under electric field. When a polarity reversal occurs in a conventional DC transmission system, the electric field at the interface is rapidly increased and even leads to breakdown. Therefore, space charge is a key property that cannot be ignored in DC cable research. This chapter finds that graphene has an inhibitory effect on injecting and accumulating of space charge in LDPE. The space charge of graphene/LDPE nanocomposites are prepared with filler contents of 0, 0.003, 0.005, 0.007 and 0.01 wt%. In addition, carrier mobility was evaluated by the depolarization process of space charge.

Space Charge Accumulation

Figure 5 shows the space charge distribution of composites with different graphene contents at different temperatures. Both ambient temperature changes and graphene content significantly affect the injection and accumulation of space charge. As shown in Figure 5 (a-1)–(d-1), most of the space charge at 0°C, 0.003, 0.005, and 0.007 wt% is concentrated near the two electrodes at 40°C, and the injected charge The amount is small, and the polarity and amount of charge remain stable under the applied voltage. The results show that space charge injection and transport are not active at low temperature and low field. As the ambient temperature increases from 40 to 80°C, the space charge injection and accumulation in the LDPE is significantly increased. Figure 5 (a-2)–(a-3) show that the polarity and the amount of space charge vary with polarization time.

As shown in Figures 5 b, c, the space charge behavior of LDPE was significantly improved when the graphene content was 0.003 and 0.005 wt% at 60 and 80 °C. However, the graphene content of 0.007 and 0.01 wt%

Figure 5. Space charge distribution with different graphene content (+30 kV/mm, 40, 60 and 80 °C), (a) 0 wt %, (b) 0.003 wt %, (c) 0.005 wt %, (d) 0.007 wt %, and (e) 0.01 wt %

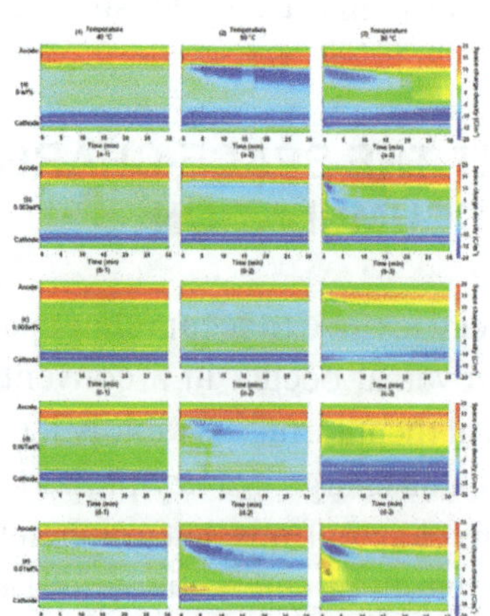

does not improve the space charge characteristics. The effects of ambient temperature and graphene content on the space charge characteristics of the composite are analyzed below.

For pure LDPE, space charge is rarely injected into the matrix at 40°C, but space charge injection and accumulation at 60 and 80°C will increase, as shown in Figure 6 (a-2)–(a-3). The space charge dynamic behavior of pure LDPE samples at 60 and 80 °C was analyzed and compared. For the pure LDPE at 60 °C shown in Figure 6, when the polarization time was 1 minute, electrons were found to be injected from the cathode. At the same time, electric charges were also observed near the anode due to space charge injection and impurity ionization. As the polarization process progresses, electrons injected from the cathode are transported to the anode and trapped by the trap, resulting in charge accumulation near the anode. In addition, the negative charge generated by ionization of impurities is trapped and accumulated near the anode, causing a sharp increase in the electric field near the anode, reducing the barrier of Schottky injection, and further injecting holes from the anode after applying the electric field for 10 minutes, such as Figure 6b shows. Figure 7 shows the recombination process between electrons

Figure 6. Space charge distribution in the neat LDPE sample (+30 kV/mm, 60 °C). The polarization time is (a) 1, (b) 5, (c) 10, (d) 20 and (e) 30 min.

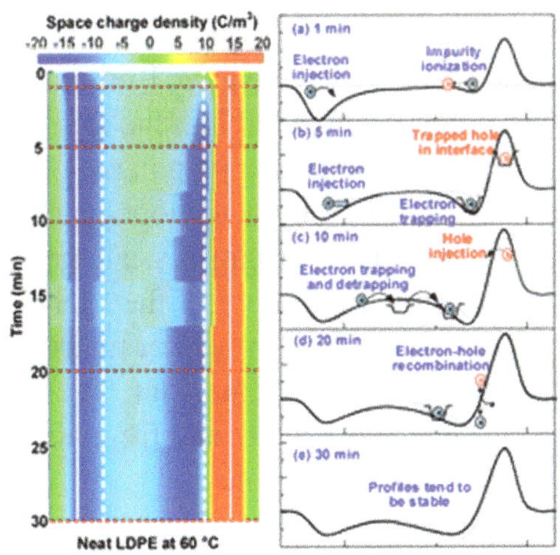

Figure 7. Space charge distribution in the neat LDPE sample (+30 kV/mm, 80 °C). The polarization time is (a) 1, (b) 5, (c) 10, (d) 20 and (e) 30 min.

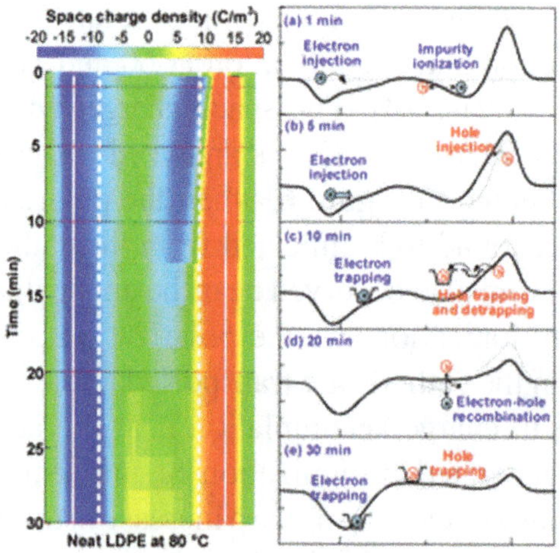

and holes occurring near the anode. The space charge distribution tends to be relatively stable within 30 minutes. The results show that an increase in ambient temperature from 40°C to 60°C has a large effect on electron injection and accumulation in LDPE samples.

When the temperature is raised to 80°C, the injection and transport of electrons (including implantation and ionization) are accelerated, so that the space charge accumulation is more severe after 1 minute of polarization. A large amount of positive charge is injected into the sample from the anode, as shown in Figure 7b. After 10 minutes of polarization, the heteropolar charge near the anode was replaced by a homopolar charge, as shown in Figure 7c. As the polarization process progresses, a large amount of positive charge continues to migrate into the interior of the LDPE, and an electron-hole recombination process occurs in the polymer matrix. After 30 minutes of polarization, positive charge became dominant. The above results show that in pure LDPE samples, the temperature rise from 60 to 80°C effectively promotes the injection and transport of holes.

In a graphene/LDPE composite with different filler contents at 40 °C, most of the same-polar charge accumulates near the electrode, except for the graphene content shown in Figure 5 (e-1). Outside of the LDPE nanocomposite, almost no space charge is injected into the matrix.

At a temperature of 60 °C, the material ionizes and accelerates the injection of electrons, and the negative charge becomes dominant. By comparing the space charge color images in Figures 5 (a-2)–(c-2), a significant effect of the addition of graphene nanoparticles on space charge injection and transport was observed. For composites with a graphene content of 0.005 wt%, the amount of charge at the location marked with a white dashed line remains constant over time, and a small amount of negative charge is believed to be produced by ionization of the impurities, as shown in Figure 5 (c). It is believed that the incorporation of a small amount of graphene nanoparticle as a filler can effectively inhibit the transport (injection and ionization) of space charge, which is caused by a large number of deep traps introduced by the polymer-graphene interface region. As the graphene content is further increased from 0.005 wt% to 0.01 wt%, the amount of space charge accumulated in the sample is significantly increased, indicating that the space charge injection and transportation process is intensified. The reasons are as follows. First, the agglomeration between adjacent fillers and the local overlap between the interface regions can weaken the influence of the deep trap sites. On the other hand, agglomerated graphene nanometers provide a conductive path for carriers in localized regions, thereby aggravating the transfer of space charge.

When the ambient temperature is further increased from 60°C to 80°C, the injection and transport of holes will be further accelerated. It can be seen from Figure 5 (a-3)–(c-3) that as the graphene filler content increases to 0.005 wt%, the depth of space charge injection is significantly reduced, which is more pronounced than electrons. For pure LDPE, in the initial stage of polarization, the heteropolar charge accumulates near the anode side, severely distorting the local electric field. As a large number of holes are injected from the anode, the same polarity charge near the anode dominates, resulting in a significant decrease in the electric field near the anode and an increase in the electric field near the cathode. The injection of space charge and local field distortion seriously affect the dielectric properties of the dielectric under DC electric field. In Figure 5 (c-3), it can be found that 0.005 wt% of the nanocomposite in the region near the two electrodes, the same polarity charge It is dominant, and the electric field remains constant, indicating that the graphene content of 0.005 wt% can effectively suppress the charge injection and transport process.

In Figure 5 (d-3), it can be found that when the content of graphene filler is 0.01 wt%, the injection and transmission process of the same polarity charge is significantly accelerated by the combination of electric field and high temperature, resulting the electric field will be significantly enhanced in the sample

The results shown above indicate that for any temperature between 40 and 80 °C, the interaction between the graphene and LDPE matrix will minimize the amount of space charge injected by 0.005 wt% of the graphene/LDPE nanocomposite, and the electric field is distorted. The weakest. When the graphene content reaches 0.01 wt%, the same polarity charge injection and transport process accelerates, which aggravates the space charge injection and accumulation in the LDPE nanocomposite.

Carrier Mobility

At the end of the polarization process, the space charge behavior of the different samples at 60°C is shown in Figure 8. The ends of the sample were short-circuited to observe the effect of graphene nano-addition on carrier

Figure 8. Space charge dissipation in LDPE composites (60 °C). (a) 0, (b) 0.003, (c) 0.005, (d) 0.007 and (e) 0.01 wt %

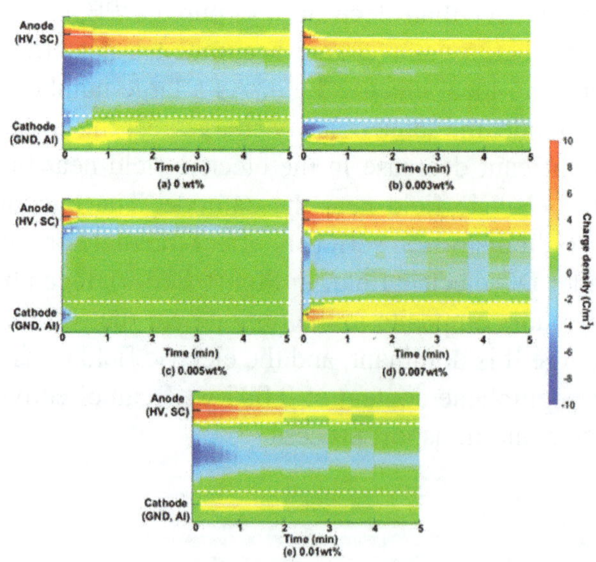

mobility. It was found that during the depolarization of the composite, the positive charge at the interface near the electrode and the negative charge accumulated in the sample body were consistent with the polarization process. When the graphene content is increased from 0 to 0.005 wt%, the injection amount and depth of the negative space charge are remarkably lowered, but when the graphene content is increased from 0.005 to 0.01 wt%, the injection amount and depth of the negative space charge are increased.At an ambient temperature of 60°C, the average charge density (q(t)) of the LDPE sample during depolarization for 15 minutes is shown in Figure 9. The carrier mobility of the nanocomposite and the corresponding trap energy and distribution As shown in Figure 10. It is found in Figure 10 that the average charge density in pure LDPE exceeds 5 C / m3 at 5 seconds and nonlinearly decays as the depolarization process proceeds. The trap involved in the depolarization process is deep, and the decay rate of the charge density decreases as the depolarization process progresses. It has also been observed that 0.003, 0.005 and 0.007 wt% of the composite material has a higher initial charge density and a slower charge decay rate than pure LDPE. As can be seen in Figure 9, 0 wt% of the remaining charge density is the largest and the remaining charge density of 0.005 wt% is the smallest before 15 minutes of depolarization time. Figure 10a shows that carrier mobility decreases as the depolarization process progresses. During the depolarization time varying from 100 to 1000 s, 0.01 wt% of carrier mobility is the largest, and 0.005 wt% of carrier mobility is

Figure 9. Relationship between charge density and depolarization time (60 °C.)

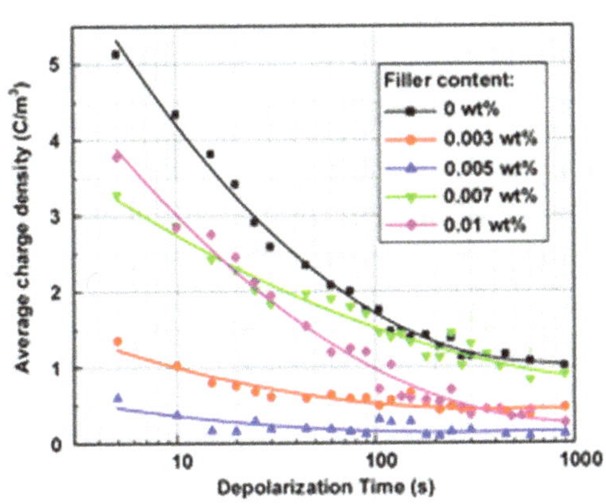

Figure 10. (a) Carrier mobility and (b) trap level (60 °C.)

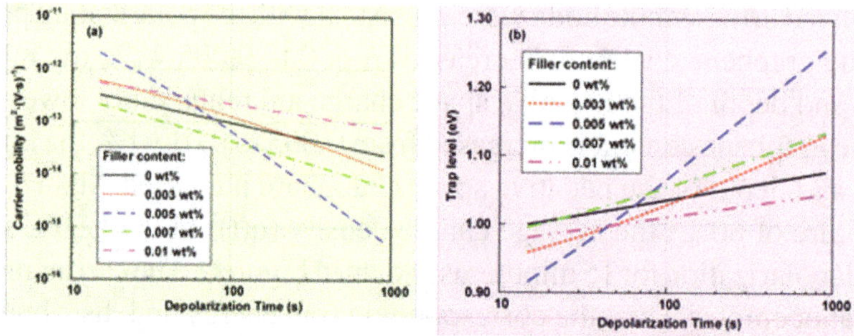

the smallest. In the first few seconds, the composite material values of 0.003 and 0.005 wt% were large, probably due to the depolarization process of the charge at the electrode interface. Figure 10b shows the relationship between trap depth and depolarization time, indicating that 0.005 wt% of the composite has a deeper trap level than the pure LDPE sample. However, as the graphene content proceeds further, the deep trap level will be further reduced. When the graphene content reaches 0.01% by weight, the carrier mobility will be higher than that of pure LDPE.

TEMPERATURE DEPENDENT TRAP LEVEL DISTRIBUTION

Trap is an important indicator of dielectric insulation performance. According to the multi-core model proposed by Tanaka, the trap is located in the bonding layer and the bonding layer of the spherical nano-substrate interface, and the charge movement needs to jump the barrier, thereby suppressing the charge transfer process. In this study, the samples prepared in the first two chapters were also used for isothermal surface potential decay (ISPD) experiments. The trap level characteristics of graphene/LDPE composites are characterized. The effects of ambient temperature and graphene addition on trap level distribution characteristics were analyzed.

Influencing Factor 1: Graphene Content

The ISPD curve of pure LDPE and composite at 70 °C is shown in Figure 11a. The initial surface potential of all samples after corona charging was found to be approximately -3800V. The ISPD curve is an exponential function of the depolarization time. As the depolarization process progresses, the decay rate of the surface potential gradually decreases. This is because the charge is designed to be more and more suitable in the process of transmission. When the graphene content is increased from 0 to 0.005 wt%, the surface potential decay rate is remarkably reduced, and when the graphene content is further increased from 0.005 to 0.01 wt%, the surface potential decay rate is gradually increased. During the polarization process, the charge accumulated on the surface of the sample may induce an electric field, which is related to the polarity of the injected charge. Under the action of the induced electric

Figure 11. Surface potential decay curves and the trap level distributions (70 °C)

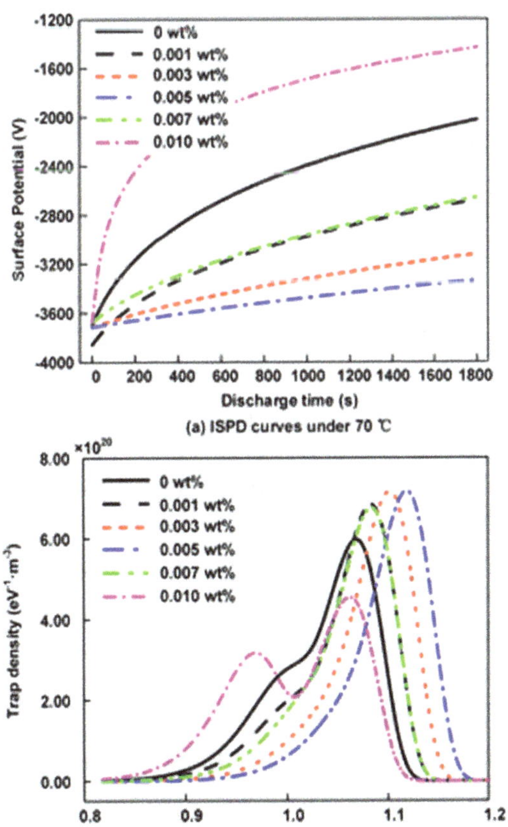

field, the surface charge is transferred to the ground electrode and undergoes the trapping and detrapping process during the transmission, resulting in a decrease in the surface potential. The surface potential of pure LDPE is -2026 V until 1800 s, while the surface potential of 0.01 wt% material is only -1433V. The surface potential of 0.005 wt% of the material is the largest, and the amplitude is -3335 V, indicating that the surface charge dissipation process is significantly inhibited.

Studies have shown that a double exponential function can be used to fit the ISPD attenuation curve. Usually, the fit curve can be expressed as:

$$V_s(t) = A_1 \exp(-t/\tau_1) + A_2 \exp(-t/\tau_2) \tag{4-1}$$

Where Vs(t) is the surface potential of the sample at time t. A1, A2, τ1 and τ2 are the fitting coefficients of the function.

From this equation, the attenuation curve can be fitted and the relationship between the trap density of the composite and the trap level at 70°C is calculated, as shown in Figure 11b.

Figure 11b shows that the trap level distribution curve is mountain-like and has two peaks, which can be understood as shallow traps and deep traps, respectively. This is based on the result of a double exponential function fit, which represents the contribution of A1exp(-t /τ1) and A2exp(-t /τ2), respectively, as shown in Figure 12. The shallow and deep trap level distribution of the composite is shown in Figure 13. It can be seen that as the graphene content increases from 0% to 0.005 wt%, the shallow trap level gradually increases and the density decreases. However, when the graphene content is further increased from 0.005% to 0.01% by weight, the shallow trap level is lowered and its total density is increased. In comparison, 0.005 wt% of LDPE composites have the highest deep trap density and energy level, that is, when the graphene content is 0.005 wt%, the composite has the highest deep trap density and the lowest shallow trap density. This will have an important impact on suppressing electron injection, affecting the capture and decapture of carriers, and effectively reducing the mean free path of electrons.

Effect of Ambient Temperature

It has also been found that the ISPD process is related to ambient temperature, presumably affecting the charge injection and transport process. Figure 14 shows the ISPD curves for composites with graphene contents of 0, 0.005

Figure 12. Trap level decomposition of neat LDPE

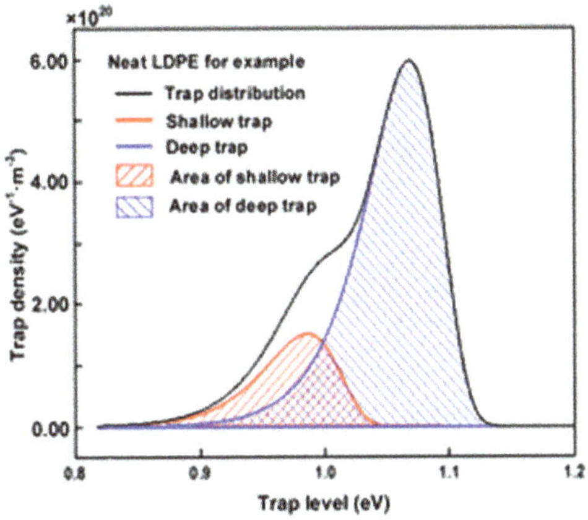

Figure 13. Shallow and deep trap level distributions of nanocomposites (70 °C.)

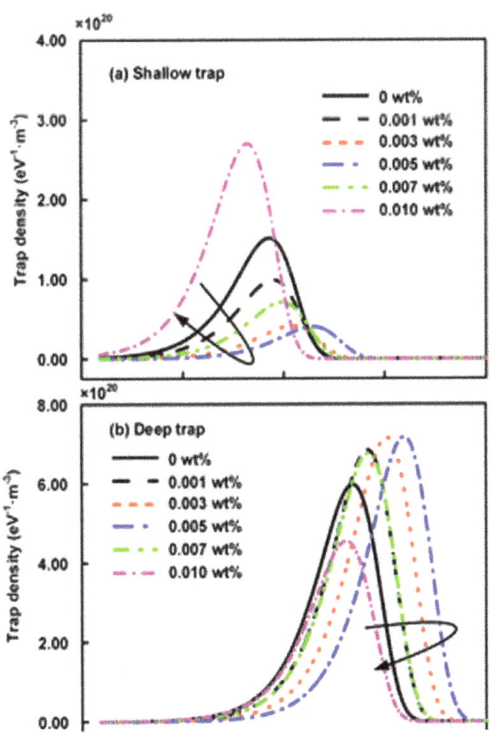

Figure 14. ISPD curves of (a) neat LDPE, (b) 0.005 wt% and (c) 0.01 wt% composites (30, 50 and 70 °C.)

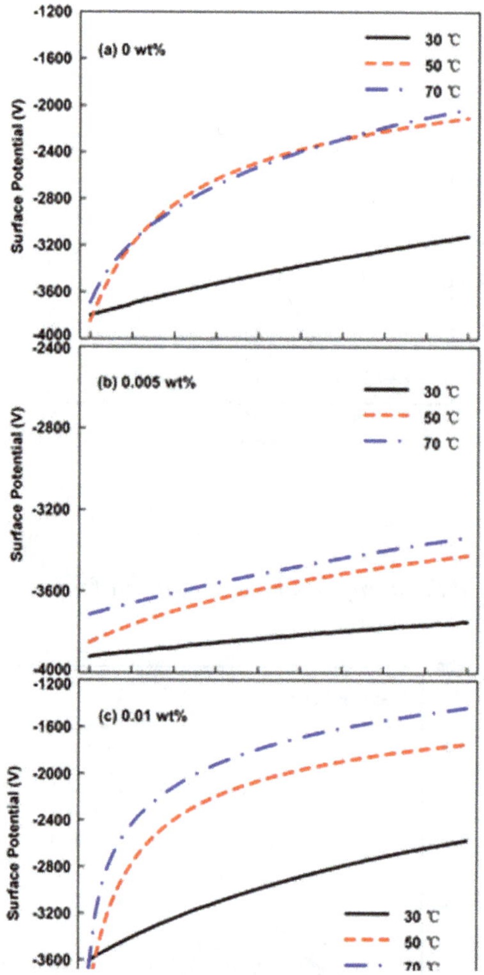

and 0.01 wt% at 30, 50 and 70°C. The surface charge decay curve was found to be very sensitive to ambient temperature. After 1800 seconds in the ISPD test, the surface potential of the pure LDPE samples decreased by -678 V and -1668 V at 30 and 70°C, respectively. Studies have shown that as the temperature increases, the thermal motion of the molecular chain becomes more active, and the trapped space charge will gain more heat, resulting in a greater probability of trapping the charge trapped in the deeper capture sites. The trap is transferred to the ground electrode. Therefore, as the ambient

temperature increases, the rate of decline of the surface potential curve of the sample increases significantly, as shown in Figure 14.

In addition, it was found that the surface potential of the composite having a graphene content of 0.005 wt% at a different temperature was much slower than that of pure LDPE and a material having a content of 0.01 wt%, which was particularly remarkable at 70 °C. After injecting a charge of 1800 s at 70°C, the surface potentials of the 0, 0.005, and 0.01 wt% samples decreased by -1668 V, -381 V, and -2238 V, respectively, indicating that more charge was trapped at 0.005. The deep trap in the wt% sample. In practical applications, the polyethylene in the HVDC cable is typically operated at high temperatures. Therefore, the charge transport behavior near the operating temperature can be improved by filling the graphene nanoparticle with a content of 0.005%.

Figure 15 shows the relationship between the trap level density of the three materials and the trap level at different temperatures. It was found that with the increase of temperature, the trap levels of these three groups of LDPE samples increased significantly, and the trap density gradually decreased. During corona charging, an external electric field drives charge injection into the surface of the sample and moves inside the polymer, causing charge to build up on the sample. Assuming that the amount of charge injected into the sample during the same charging time is constant, then as the depolarization process proceeds, higher temperatures will cause charge carriers to collapse in deeper traps, accelerating the surface potential. attenuation. That is, at 30°C, the space charge of the deep trap is difficult to capture, and the deep trap density is almost zero. As the temperature increases, the trap of the LDPE sample becomes deeper and deeper, as shown in Figure 15.

Mechanism of Graphene Nanofiller

Figure 16 shows the deep trap and shallow trap levels for composites at 30°, 50°, and 70°C, and the deep trap density at 70°C. The deep traps and shallow trap levels of LDPE were found to be significantly affected by temperature. The reason is that at lower temperatures, most of the electrons do not have enough energy in the conduction band, and a small portion of the electrons with sufficient energy are quickly trapped by the trap. As the temperature increases, the thermal motion of the charge becomes intense, accelerating the electron capture and transport process, resulting in an increase in the mean free path of the electron.

Figure 15. Trap level distributions of (a) neat LDPE, (b) 0.005 wt% and (c) 0.01 wt% nanocomposites (30, 50 and 70 °C.)

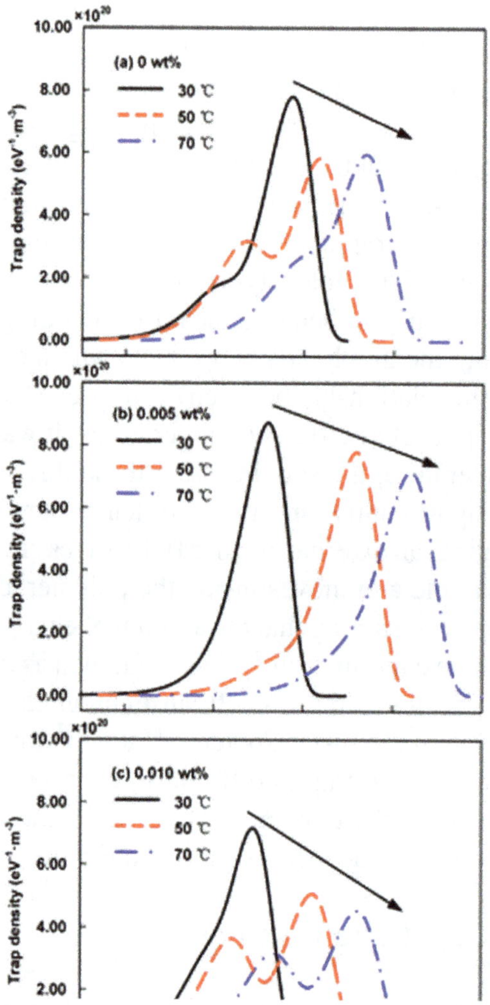

The above experimental results show that nanographene introduces a deep trap into the material, which significantly inhibits the surface charge transport and dissipation in the composite, especially the sample with a content of 0.005 wt%. Figure 17 is a schematic diagram of the trap affecting the charge transport process. According to the multi-core model proposed by Tanaka, a large number of deep traps are distributed in the inner layer of the interface region between the graphene filler and the LDPE matrix. The outer layer, the loose layer, contains many shallow traps. The large number of

Figure 16. (a) Shallow and (b) deep trap level (30, 50 and 70 °C), (c) deep and shallow trap densities under ambient temperature of 70 °C

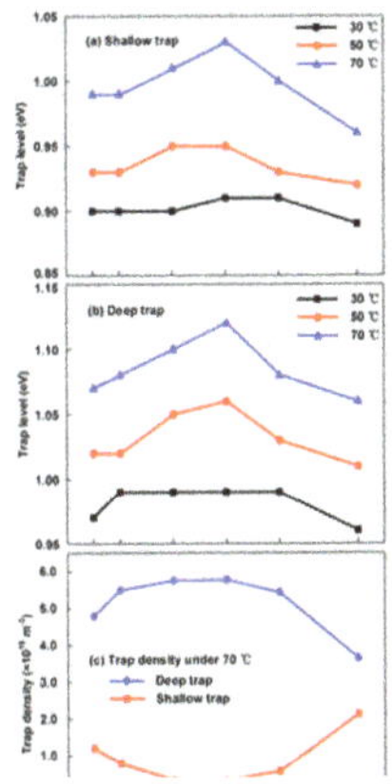

Figure 17. Deep trap modulated charge carrier transport

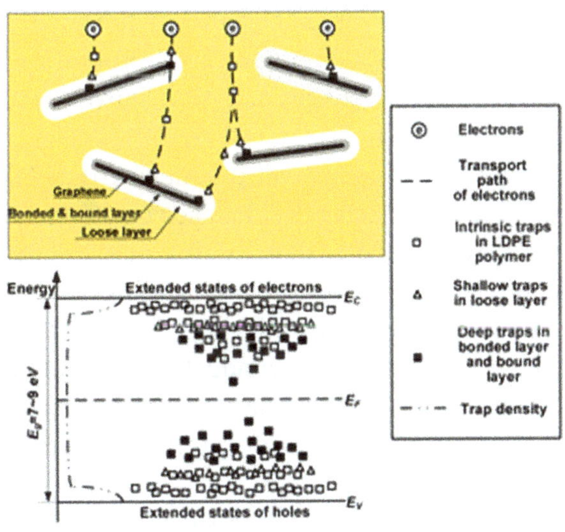

deep traps introduced by well-dispersed graphene interact with the intrinsic trap sites in the LDPE matrix, resulting in an increase in the density of deep traps in the composite. Deep traps in the interface region reduce the rate of electron desorption and inhibit charge injection from the electrode into the polymer matrix.

It is also found in Figures 17b and c that when the filler content is further increased from 0.005 wt% to 0.01 wt%, the deep trap level and density decrease, presumably due to the overlap between the loose layers of adjacent graphene nanofillers. of. The overlap causes the influence of deep traps in the interface region to be weakened, resulting in a decrease in deep trap density. Since graphene has a high electrical conductivity, it provides a low resistance path for charge when it agglomerates, thereby accelerating the charge transport process.

CONCLUSION

High-voltage DC cables are an important development direction for future grid systems. DC cables suitable for high-voltage systems are required to have low conductivity, high breakdown field strength, and less space charge accumulation. Nanocomposites have become an important development direction of high-voltage DC cables because of these advantages. In this chapter, the research background and research progress of graphene/low density polyethylene nanocomposites are first introduced. Then, several composite materials of 0, 0.001, 0.003, 0.005, 0.01, 0.1 wt% were prepared by adding different amounts of nano-scale graphene to the LDPE. By studying its conductivity, breakdown electric field, space charge and trap properties, the following laws were discovered:

1. The conductivity of composites after adding low concentration of graphene nanometers is significantly reduced, which is caused by the barrier of increasing the charge of traps. The conductivity of the material after the addition of high concentration of graphene nano-nano is deteriorated, because the high-conductivity graphene forms a conductive path in the LDPE, allowing the charge to flow through the channel.
2. The electrical properties of the composite are affected by the graphene content and the ambient temperature. The physical laws of these two influencing factors are explained by establishing correlation models. The interfacial region between graphene and the polymer introduces many deep traps in the forbidden band of the nanocomposite, which can reduce the charge trapping process and suppress the accumulation of space charge.

REFERENCES

Fim, F. D. C., Basso, N. R. S., Graebin, A. P., Azambuja, D. S., & Galland, G. B. (2013). Thermal, electrical, and mechanical properties of polyethylene–graphene nanocomposites obtained by in situ polymerization. *Journal of Applied Polymer Science*, *128*(5).

Gaska, K., Xu, X. D., Gubanski, S. M., & Kádár, R. (2017). Electrical, mechanical, and thermal properties of LDPE graphene nanoplatelets composites produced by means of melt extrusion process. *Polymer*, *9*(1), 11–22. doi:10.3390/polym9010011 PMID:30970688

Gray, A. P. (1970). Polymer crystallinity determinations by dsc. *Thermochimica Acta*, *1*(6), 563–579. doi:10.1016/0040-6031(70)80008-9

Huang, X., Jiang, P., & Yin, Y. (2009). Nanoparticle surface modification induced space charge suppression in linear low density polyethylene. *Applied Physics Letters*, *95*(24), 242905, 242905–3. doi:10.1063/1.3275732

Jung, J., Kim, J., Uhm, Y. R., Jeon, J. K., & Rhee, C. K. (2010). Preparations and thermal properties of micro- and nano-bn dispersed hdpe composites. *Thermochimica Acta*, *499*(s 1–2), 8–14.

Krivda, A., Tanaka, T., Frechette, M., Castellon, J., Fabiani, D., Montanari, G. C., ... Anglhuber, M. (2017). Characterization of epoxy microcomposite and nanocomposite materials for power engineering applications. *IEEE Electrical Insulation Magazine*, *28*(2), 38–51. doi:10.1109/MEI.2012.6159180

Kumara, J. R. S. S., Serdyuk, Y. V., & Gubanski, S. M. (2016). Surface potential decay on ldpe and ldpe/al2o3 nano-composites: Measurements and modeling. *IEEE Transactions on Dielectrics and Electrical Insulation*, *23*(6), 3466–3475. doi:10.1109/TDEI.2016.005663

Lan, L., Wu, J., Yin, Y., Li, X., & Li, Z. (2014). Effect of temperature on space charge trapping and conduction in cross-linked polyethylene. *IEEE Transactions on Dielectrics and Electrical Insulation*, *21*(4), 1784–1791. doi:10.1109/TDEI.2014.004261

Lau, K. Y., Vaughan, A. S., Chen, G., Hosier, I. L., & Holt, A. F. (2013). On the dielectric response of silica-based polyethylene nanocomposites. *Journal of Physics. D, Applied Physics*, *46*(9), 095303. doi:10.1088/0022-3727/46/9/095303

Lewis, T. J. (1994). Nanometric dielectrics. *IEEE Transactions on Dielectrics and Electrical Insulation, 1*(5), 812-825.

Lewis, T. J. (2004). Interfaces are the dominant feature of dielectrics at the nanometric level. *IEEE Transactions on Dielectrics and Electrical Insulation, 11*(5), 739-753.

Li, M., Gao, C., Hu, H., & Zhao, Z. (2013). Electrical conductivity of thermally reduced graphene oxide/polymer composites with a segregated structure. *Carbon, 65*, 371–373.

Li, S., Zhu, Y., Min, D., & Chen, G. (2016). Space charge modulated electrical breakdown. *Scientific Reports, 6*(32588). doi:10.1038rep32588 PMID:27599577

Li, X., Xu, M., Zhang, K., Xie, D., Cao, X., & Liu, X. (2014). Influence of organic intercalants on the morphology and dielectric properties of xlpe/montmorillonite nanocomposite dielectrics. *IEEE Transactions on Dielectrics and Electrical Insulation, 21*(4), 1705–1717. doi:10.1109/TDEI.2014.004317

Li, Z. F., Zhang, H., Liu, Q., Sun, L., & Xie, J. (2013). Fabrication of high-surface-area graphene/polyaniline nanocomposites and their application in supercapacitors. *ACS Applied Materials & Interfaces, 5*(7), 2685–2691. doi:10.1021/am4001634 PMID:23480549

Montanari, G. C. (2011). Bringing an insulation to failure: the role of space charge. *IEEE Transactions on Dielectrics and Electrical Insulation, 18*(2).

Nelson, J. K. (2002). Toward an understanding of nanometric dielectrics. *2002 Annual Report CEIDP*.

Nelson, J. K., & Fothergill, J. C. (2004). Internal charge behaviour of nanocomposites. *Nanotechnology, 15*(5), 586–595. doi:10.1088/0957-4484/15/5/032

Nelson, J. K., & Hu, Y. (2005). Nanocomposite dielectrics—Properties and implications. *Journal of Physics. D, Applied Physics, 38*(2), 213–222. doi:10.1088/0022-3727/38/2/005

Peng, S., Zeng, Q., Yang, X., Hu, J., Qiu, X., & He, J. (2016). Local dielectric property detection of the interface between nanoparticle and polymer in nanocomposite dielectrics. *Scientific Reports, 6*(1), 38978. doi:10.1038rep38978 PMID:27958347

Rafiee, M. A., Rafiee, J., Wang, Z., Song, H., Yu, Z. Z., & Koratkar, N. (2009). Enhanced mechanical properties of nanocomposites at low graphene content. *ACS Nano*, *3*(12), 3884–3890. doi:10.1021/nn9010472 PMID:19957928

Tanaka, T., Bulinski, A., Castellon, J., Frechette, M., & Han, S. J. (2011). Dielectric properties of xlpe/sio2 nanocomposites based on cigre wg d1.24 cooperative test results. *IEEE Transactions on Dielectrics and Electrical Insulation*, *18*(5), 1482–1517. doi:10.1109/TDEI.2011.6032819

Tanaka, T., & Imai, T. (2013). Advances in nanodielectric materials over the past 50 years. *IEEE Electrical Insulation Magazine*, *29*(1), 10–23. doi:10.1109/MEI.2013.6410535

Tanaka, T., Kozako, M., Fuse, N., & Ohki, Y. (2005). Proposal of a multi-core model for polymer nanocomposite dielectrics. *IEEE Transactions on Dielectrics and Electrical Insulation*, *12*(4), 669–681. doi:10.1109/TDEI.2005.1511092

Tanaka, T., Montanari, G. C., & Mulhaupt, R. (2004). Polymer nanocomposites as dielectrics and electrical insulation-perspectives for processing technologies, material characterization and future applications. *IEEE Transactions on Dielectrics and Electrical Insulation*, *11*(5).

Wang, Y., Li, G., Wu, J., & Yin, Y. (2016). Effect of temperature on space charge detrapping and periodic grounded dc tree in cross-linked polyethylene. *IEEE Transactions on Dielectrics and Electrical Insulation*, *23*(6), 3704–3711. doi:10.1109/TDEI.2016.005986

Wu, K., Wang, Y., Wang, X., Fu, M., & Hou, S. (2017). Effect of space charge in the aging law of cross-linked polyethylene materials for high voltage dc cables. *IEEE Electrical Insulation Magazine*, *33*(4), 53–59. doi:10.1109/MEI.2017.7956633

Zha, J. W., Song, H. T., Dang, Z. M., Shi, C. Y., & Bai, J. (2008). Mechanism analysis of improved corona-resistant characteristic in polyimide/tio2 nanohybrid films. *Applied Physics Letters*, *93*(19), 192911–0. doi:10.1063/1.3025408

Chapter 6
Characteristics and Suppression of Space Charge in Polyethylene

ABSTRACT

In HVDC transmission systems, the space charge accumulation ofpolyethylene (PE) insulation is a major problem that threatens the safe and stable operation of cables. In this chapter, nanoparticles and voltage stabilizers are used to inhibit space charge in PE, which has excellent compatibility with PE. To study the thermal, mechanical, and electrical properties of the samples, differential scanning calorimetry (DSC) testing, tensile testing, breakdown, and conductivity property were measured separately. Besides, the space charge behavior based on the PEA method was studied, and the carrier mobility was calculated by the space charge depolarization process. The experimental results indicate that PE modified by graphene oxide (GO) nanoparticles and the voltage stabilizers demonstrate the suppression of space charge accumulation in PE insulation, which has less space charge accumulation than pure PE. The results show that graphene oxide and the preferred stabilizer have broad prospects in HVDC cable applications.

DOI: 10.4018/978-1-5225-8885-6.ch006

INTRODUCTION

Cross-linked polyethylene (XLPE) is a significant insulation used in high-voltage cable, which plays a role in HVDC transmission (Bjorlow-Larsen, 2000). Yet, space charge accumulation problem in the XLPE may be caused by-products during the manufacturing process, leading to the insulation aging and greatly limiting the application of XLPE in HVDC cables (Montanari, 2005). Since its first introduction, nanocomposites have given rise to a new research frenzy as a recognized method for solving the space charge accumulation in polymers (Lewis, 1994). A number of studies have been conducted on methods of adding nano-particles into LDPE matrices, which have been proved to improve the space charge characteristics. For the LDPE/MgO composites, the doping of the MgO nanoparticles is effective for the space charge suppression under the stretching conditions (Wang, 2016). Moreover, studies have shown that by adding an appropriate amount of MgO nanoparticles to polyethylene, space charge can be suppressed. Besides, the DC breakdown strength of the composite can be improved (Peng, 2015). The inhibition of space charge in the nanocomposites may be due to the interface between the filler and the matrix, based on which a model of multi core has been put forward, helping to explain existed research results and has been extensively recognized. (Tanaka, 2005). The nanoparticles can solve the problem of space charge accumulation of nanocomposites to some extent. Nevertheless, the amount of additives is relatively high, which will lead to agglomeration of nanoparticles and may cause unpredictable defects in composite materials. Therefore, there is an urgent need to develop new nano-fillers to achieve the purpose of inhibiting the space charge in the polymer by less added content.

Graphene oxide (GO) is one of the most major derivatives of the graphene, which has a single layer structure with a thickness as only one atom. The specific surface area of GO is large, which makes small doping amount possible. GO also possess well properties due to the destruction of its sp2 bonding network (Novoselov, 1998). These excellent properties have made GO a widespread concern since its discovery and an ideal filler for the fabrication of nano dielectric materials (Novoselov, 2004). The electrical and mechanical properties of GO and GNP doped polyurethane nanocomposites are superior to those of pure material (Pokharel, 2015). In addition, research showed that GO-GNF/LDPE nanocomposites can inhibit the packet-like charges (Kim, 2013). However, little work has been done on the extremely low content of GO nanoparticles to suppress spatial impurities in LDPE and its mechanism analysis. In addition, there are few studies on other electrical properties of LDPE/GO nanocomposites, such as DC conductivity.

Based on extensive research, voltage stabilizers are believed to be effective in improving the performance of polymers. In the late 1960s, the voltage stabilizers has been proposed and its compatibility with polymer matrices is poor, and compatibility is quite important for the long service life and stable performance of insulating materials, so it has not attracted much attention (Gross, 1967). Voltage stabilizers with higher solubility are prepared by attaching an alkyl side chain to an aromatic nucleus (Martinotto, 2001). There are various kinds of voltage stabilizers. Aromatic ketones and diketones, such as acetophenone, benzophenone and benzyl, can play a positive role in improving the electrical properties of materials (Jarvid, 2014). Also, in the application of insulation in high voltage power cable, fullerenes and their derivatives are used as stabilizers (Jarvid, 2015). Studies have shown that voltage stabilizers with lower ionization potential are more effective for suppressing electric trees (Jarvid, 2014; Zhang, 2014). The main purpose of the existing research is to suppress the electric tree. There are few studies on the electrical properties of the voltage stabilizer under direct current condition, especially the effect of the stabilizer on the space charge is little known.

In this chapter, LDPE / GO nanocomposites were prepared with GO filler contents of 0, 0.001, 0.005, 0.01, 0.05, and 0.1 wt%, respectively. There were three different voltage-stabilizer (4,4'-Difluorobenzophenone, 4,4'-Dihydroxybenzophenone, and 4,4'-Bis (dimethyl amino) benzyl, were purchased from J&K Scientific Ltd., denoted by A, B, and C, respectively). XLPE composite samples with a filler content of 0.5% by weight were prepared. The relative dielectric constant, DC conductivity, space charge distribution, trap characteristics and DC breakdown property were tested and the experimental results were discussed. It can be known that the addition of GO nanoparticles and stabilizers have an effect on the electrical properties of PE, and its effect on space charge may be related to changes in trap levels.

EFFECT OF GRAPHENE OXIDE PARTICLES ON SPACE CHARGE ACCUMULATION

Relative Permittivity and DC Conductivity

As shown in Figure 1, the relative dielectric constant of LDPE / GO nanocomposites at 298 K varies with GO content (Du, 2018). The filler content is increased to 0.01 wt%, and the relative dielectric constant of the composite material is lowered and lower than that of the pure material. However, as the filler content continues to increase, the relative dielectric constant increases. There are some other similar findings (Singha, 2008). Since the dielectric constant of the GO particles is higher than the dielectric constant of the LDPE matrix, the dielectric constant of the composite increases. When the GO content is low, the dielectric constant of the nanocomposite is low, which may be attributed to the hindrance of the mobility of the polyethylene chain due to the interaction process between the surface of the nanoparticle and the polymer chain (Roy, 2007). Therefore, the polarization process of most LDPE / GO nanocomposites is limited.

Figure 1. Relationship between the relative permittivity and the frequency with the different GO filler content

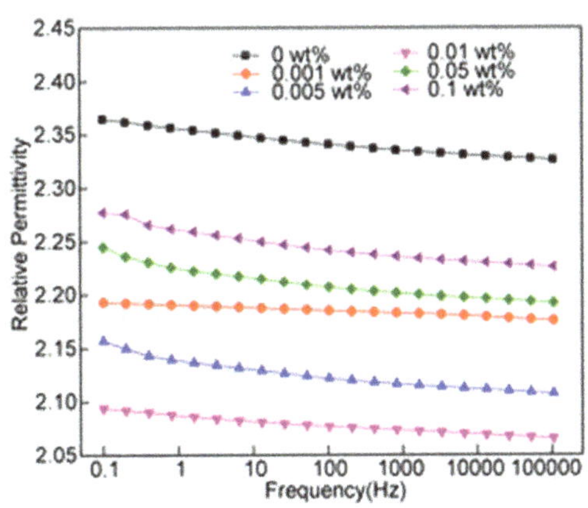

Figure 2 shows the relationship of DC conductivity and polarization time of LDPE/GO nanocomposites. From the data it can be seen the conductivity decreases with the polarization time. At a GO content of 0.01 wt%, the charge current of the composite during the first 100 s will decrease significantly. After that, the charging current approaches steady state and the decay rate is relatively slow. As the polarization time lengthens, the polarization current gradually stabilizes and the conduction current becomes dominant, thus the average conductivity of the last 100 s is regarded as the DC conductivity. For GO content of 0, 0.001, 0.005, 0.01, 0.05 and 0.1 wt% respectively, the conductivity of nanocomposites are 1.47×10-14, 1.09×10-15, 2.47×10-16, 1.77×10-16, 4.51×10-16 and 2.35×10-15 S/m. The 0.01% by weight composite material has a much lower direct current conductivity than pure LDPE. GO has a high conductivity under a DC electric field, as described in Section 2.1. According to the percolation theory, when the filler content is low, it can be uniformly dispersed in the polymer matrix. As shown in Figures 1b and 1c, the average distance between adjacent filler particles is several tens of microns, that is, the filler content below 0.01 wt% is well below the percolation threshold. The original trap site at the interface between the LDPE crystalline phase and the amorphous phasecaptures carriers, and there is no conductive path for carriers formed by the entire LDPE / GO composite. Moreover, due to the large specific surface area of GO, a large number of interaction regions are at the interface between the GO and LDPE polymers. The polymer

Figure 2. DC conductivity of LDPE/GO nanocomposites as a function of polarization time

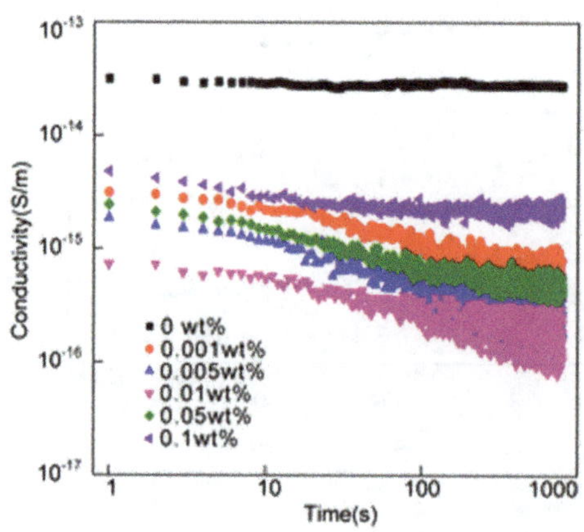

and filler interaction zone captures charge carriers, thereby inhibiting the transport of charge carriers. When the GO filler content is more than 0.01% by weight, the direct current conductivity begins to increase. The average distance between adjacent GO fillers gradually decreases as the GO content increases. The interaction zones of the polymer and the filler overlap each other, providing a path for the carriers and accelerating their local transport, thereby promoting an increase of DC conductivity.

Suppressing Space Charge

Figures 3 and 4 show the dynamic space charge distribution and electric field for samples with different GO filler contents, respectively. Since LDPE is a semi-crystalline material, the injected charge can be trapped by traps at the interface between the crystalline phase and the amorphous phase (Lewis, 2002). As the polarization time increases, the injected charge moves from the cathode to the anode. The same polarity charge accumulates near the anode, causing significant electric field distortion in the vicinity of the electrode.

Figure 3a shows the evolution of space charge in pure LDPE. It can be seen that the space charge shows an increasing trend as the polarization time increases. After polarization, a significant accumulation of the same polarity charge can be observed near the anode. The maximum value of the space charge density is higher than 7 C/m^3, which will distort the local electric field

Figure 3. Space charge density of LDPE/GO nanocomposites: (a) 0 wt%, (b) 0.001 wt%, (c) 0.005 wt%, (d) 0.01 wt%, (e) 0.05 wt%, and (f) 0.1 wt%

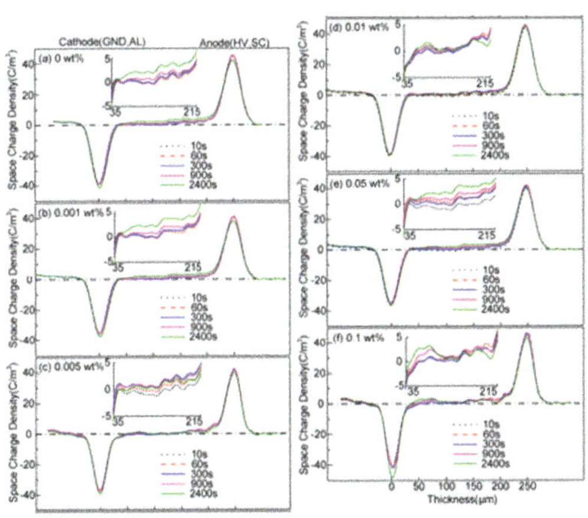

near the electrode (Montanari, 2011). At the same time, as shown in Figure 4a, the electric field strength near the cathode at the end of the polarization reaches -62.47 kV/mm.

Comparing the results of space charge distribution in Figure 3 a, b, c and d, the relationship between the content of different GO fillers and the accumulation of space charge is analyzed. In most samples, the accumulated space charge decreases significantly as the GO filler increases. When the polarization time is long, only a small amount of the same polarity charge is accumulated in a narrow region near the electrode, and the amount of space charge in the body is substantially unchanged. After 2400 s of polarization, the maximum space charge density for composites with filler contents of 0.001, 0.005 and 0.01 wt% was 6.51, 4.34 and 3.51 C/m^3, respectively, which was significantly reduced compared to pure LDPE, and suppressed to some extent. The electric field is distorted. At the same time, the maximum electric field around the cathode is -62.11, -58.95 and -55.79 kV/mm, both lower than the pure LDPE material. The inhibition of charge carriers by the interaction zone in the composite material can also explain its inhibition of space charge, which is consistent with the experimental results of the DC conductivity test above.

As shown in Figures 3e and f, the space charge accumulation is exacerbated when the GO filler content exceeds 0.01% by weight. Significant homogenous charge accumulation can be seen in the sample. After polarization, the

Figure 4. Electric field of LDPE/GO nanocomposites: (a) 0 wt%, (b) 0.001 wt%, (c) 0.005 wt%, (d) 0.01 wt%, (e) 0.05 wt%, and (f) 0.1 wt%

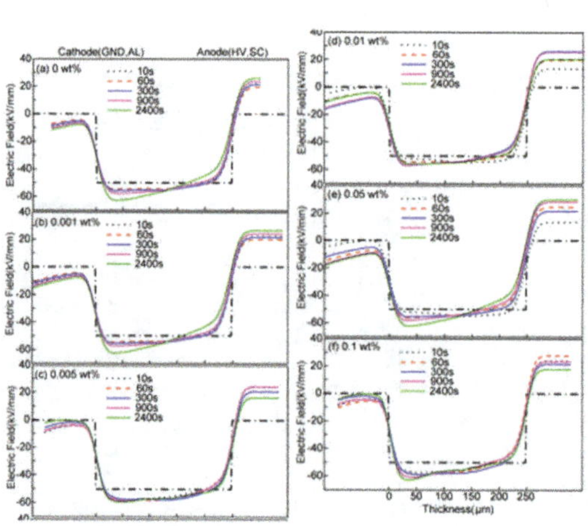

maximum space charge density of 0.05 and 0.1 wt% nanocomposites exceeded 5 and 6 C/m^3, respectively. The interaction zone of the polymer and the filler overlaps, which provides a low resistance path for the electrons while reducing the barrier between the interchain states. The transmission of electrons accelerates, eventually resulting in a large amount of space charge being injected into the sample. Therefore, as shown in Figures 4e and f, very significant electric field distortion occurs near the anode. For 0.05 and 0.1 wt% LDPE/GO samples, the maximum electric field around the cathode reached -61.61 and -62.71 kV/mm at 2400 s.

Trap Level Distribution

Based on the above experimental results, the space charge accumulation of LDPE composite containing 0.01 wt% GO is lower than that of pure LDPE, and the degree of electric field distortion is alleviated. The method of adding nanofillers can effectively suppress the carrier transport process. Studies have shown that the space charge accumulation of polymers has a great correlation with the trap level distribution (Tanaka, 2005). As shown in Figure 5, the relationship between different GO contents and trap level distribution was measured by the IDC method. The trap level of the sample is in the range of 0.8~1.05 eV, and the peak value is about 0.98 eV, which

Figure 5. Trap level distributions of LDPE/GO nanocomposites

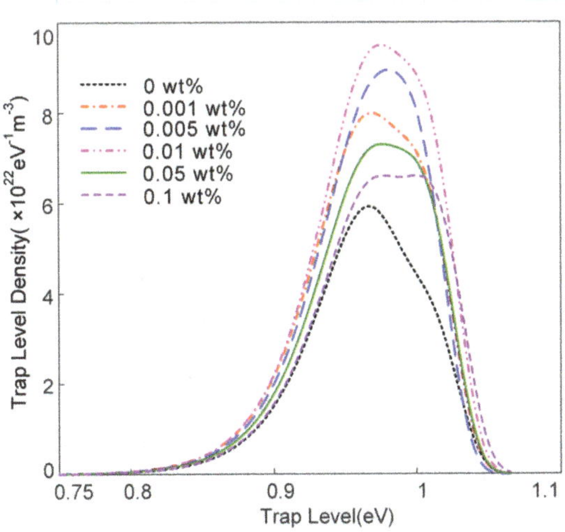

can be considered as a deep trap in LDPE (Wang, 1998). The paper (Ieda, 1984) specifies the deep trap level in LDPE as 0.92 eV, which is basically consistent with this paper. The results show that the deep well level density increases as the content of GO particles increases. However, when the GO filler content is higher than 0.01 wt% and further increased, the deep well level density is decreased.

When the content of GO nanoparticles is low, the dispensability is good. A large number of deep carrier traps were introduced that interacted with the original trap sites in the LDPE matrix, causing a significant increase in deep trap density in the composite. The deep trap introduced can capture holes and electrons to suppress the transport process of charge carriers, and this plays a role in suppressing the space charge in LDPE. At the same mass fraction, GO has a large specific surface area compared to other spherical nanoparticles, which enables it to provide deeper traps and thus has broad application prospects for improving the electrical properties of polymers. After adding 0.01 wt% GO to LDPE, the deep trap in the composite affects charge carrier transport and suppresses space charge injection. The volume fraction of the interfacial region of the polymer and filler continues to increase as the filler content continues to increase, while the average distance between adjacent GO fillers gradually decreases to the submicron or nanometer scale. Thus, there is a phenomenon in which the interaction regions of the polymer and the filler overlap each other. When the distance between adjacent particles is sufficiently thin, a large number of carriers there will jump through the thin gap through thermally activated field enhancement. That is to say, this may provide a low resistance path for electrons and holes, weakening the interaction zone, and thus promoting local transport of carriers (Li, 2017). The above can explain the phenomenon of space charge injection and cumulative increase.

Electrical Field Distortion and DC Breakdown Strength

Figure 6 shows the electric field distortion coefficient, which is calculated by the method in the paper (B.X.Du, 2016), in order to more clearly reflect the change of the electric field. When the GO content in the nanocomposite is 0.01 wt%, the electric field distribution in the LDPE is improved significantly, and the distortion coefficient of the LDPE is greatly reduced.

As shown in Figure 7, the different GO filler content affects the DC breakdown strength of the composite, which is analyzed by the Weibull distribution. It can be seen that when the GO contents are 0, 0.001, 0.005, 0.01,

Figure 6. Electrical field distortion factors of LDPE/GO nanocomposites after 40 mins polariton

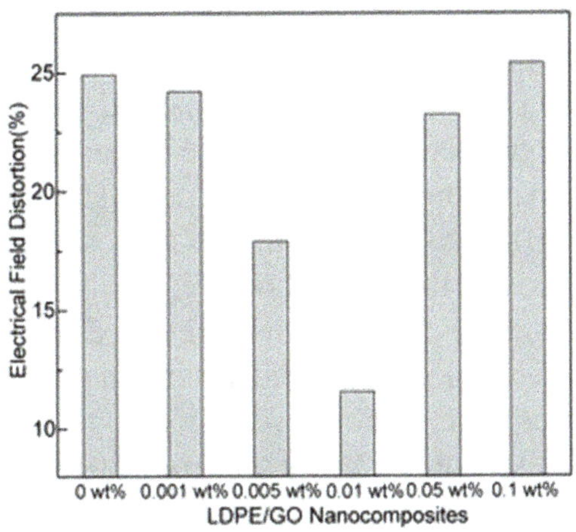

Figure 7. Relationship between the Weibull probability and the DC breakdown strength filed with the different GO filler content

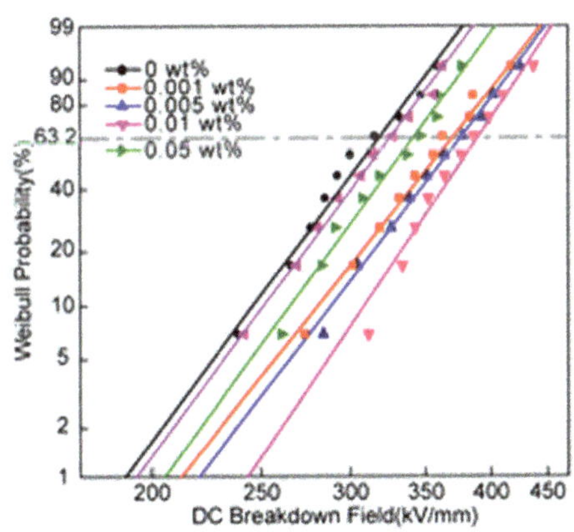

0.05, and 0.1 wt%, respectively, the corresponding composite materials have breakdown strengths of 318.2, 368.3, 376.4, 389.6, 325.3, and 337.9 kV/mm. The DC breakdown strength of the material is related to the DC conductivity and electric field distribution. The sample with lower DC conductivity and uniform electric field has higher DC breakdown strength. For a composite material having a GO content of 0.01 wt%, the DC breakdown strength is improved because the polymer and nanoparticle interaction region suppresses charge carrier transport and improves electric field distribution. The GO filler content is greater than 0.01 wt% and continues to increase, and the transport of charge carriers through the amorphous region is accelerated, thereby increasing the DC conductivity. Moreover, severe electric field distortion is caused by space charge accumulation, which may also result in partial discharges in the amorphous region, even decomposition of materials. In summary, the GO filler content is increased from 0.01 wt% to 0.1 wt%, and the breakdown strength of the composite material is lowered.

EFFECT OF VOLTAGE STABILIZERS ON THE SPACE CHARGE BEHAVIOR

Sample Characteristics

As is shown in Figure 8, different stabilizing stabilizers affect the melting and crystallization curves of XLPE (Du, 2019). The crystallinity X is obtained by integrating the DSC curve as shown in the following formula (Gray, 1970):

$$X = \frac{\Delta H_m}{\Delta H_0} \times 100\% \tag{3-1}$$

where ΔH_0 represents the melting enthalpy of XLPE fully crystallized (100%) and generally $\Delta H_0 = 290$ J/g (Jarvid, 2014). ΔH_m represents the melting enthalpy of XLPE tested. Table 1 shows the melting enthalpy, melting temperature and calculated crystallinity of XLPE before and after modification. It can be seen that after the addition of the voltage stabilizer, the melting temperature of the XLPE composite increases and the crystallinity decreases. Since the degree of change in melting temperature and crystallinity is very small, it is

Figure 8. DSC curves of (a) melting and (b) crystallization of the voltage-stabilizer modified XLPEs

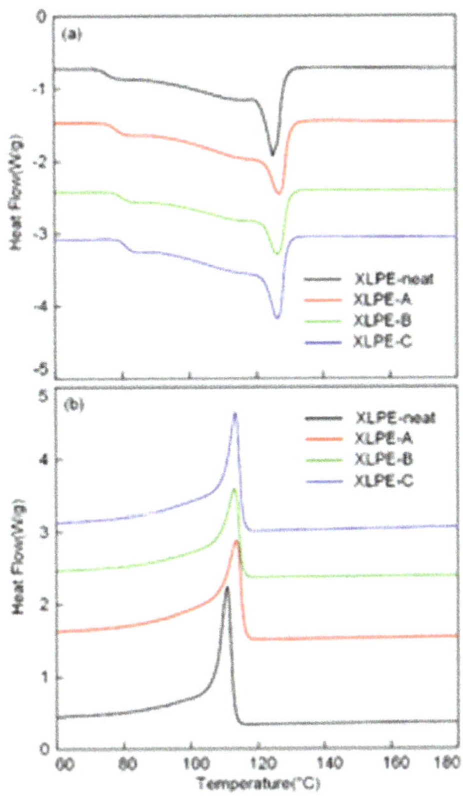

Table 1. Crystallinity and peak melting temperature of the voltage-stabilizer modified XLPEs

Samples	H_m(J/g)	Crystallinity (%)	T_m(°C)
XLPE	101.7	35.07%	125.13
XLPE-A	100.4	34.62%	126.88
XLPE-B	98.2	33.86%	126.37
XLPE-C	101.2	34.90%	126.27

considered that the stabilizer does not have much influence on the thermal properties of XLPE. This is consistent with the results obtained from previous studies (Wutzel, 2015).

As shown in Figures 9a and b, in order to study the effect of stabilizer on the mechanical properties of XLPE, the tensile properties of XLPE samples before and after stabilizer modification were compared. After the addition of the voltage stabilizer, the tensile behavior of the XLPE composite is similar to that of pure XLPE. XLPE-A containing A is even slightly better than XLPE in mechanical properties.

Figure 9. Tensile behaviors of the different voltage-stabilizer modified XLPEs: (a) stress-strain curves; (b) breaking elongation and tensile strength

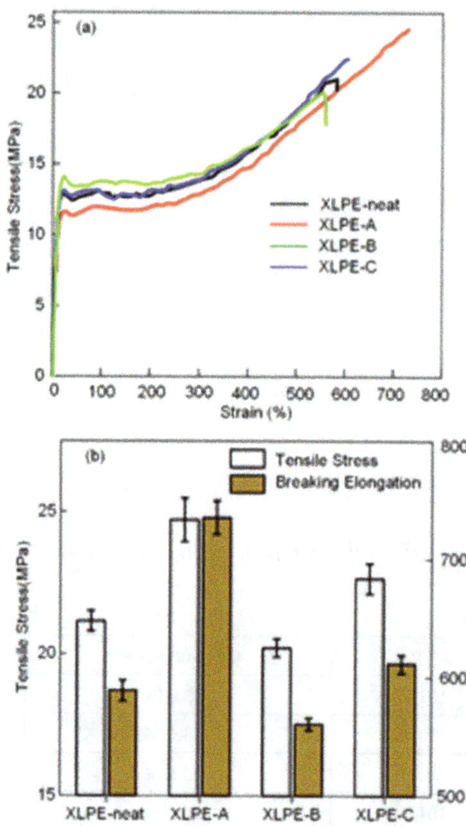

According to the above thermodynamic and mechanical experiments, it can be concluded that the stabilizer used in this study has good compatibility with XLPE, and its influence on the microstructure of the substrate is negligible, which is a necessary condition for the production of XLPE insulation materials.

Figure 10a shows the DC charging current with different polarization stabilized XLPE composites as a function of polarization time under an electric field of 10 kV/mm. As the polarization time increases, the charging current of the sample gradually decreases, and there is no significant difference between the differentsamples. The charging current consists of a polarization current, a displacement current, and a conduction current. During the first 100 s, the sample charging current dropped significantly. After the quasi-steady

Figure 10. (a) Charging current as a function of polarization time and (b) DC conductivity of different voltage-stabilizer modified XLPEs

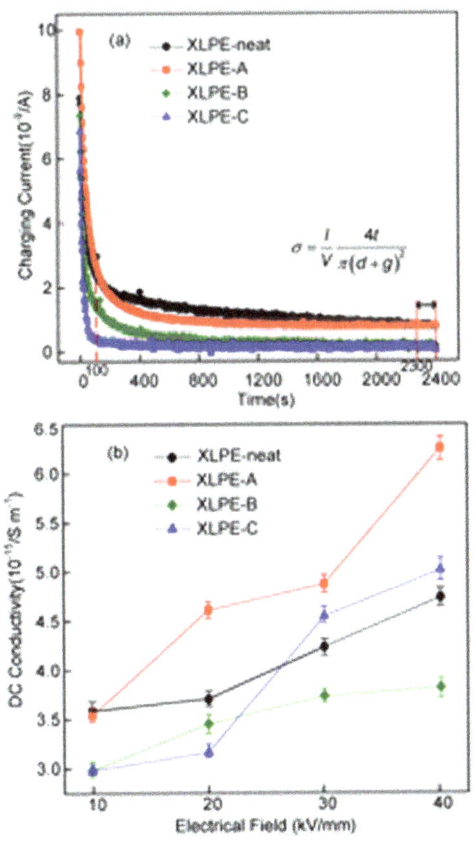

state is reached, the charging current continues to decay at a lower rate. That is, the polarization current and the displacement current gradually enter a steady state, and the conduction current eventually dominates. Therefore, the average value of the last 100 s of charging current is used as the basis for the calculation of DC conductivity. Figure 10b shows that the DC conductivity of all samples increased slightly with increasing electric field and the values were all of the same order of magnitude. It can be shown that the regulator in this experiment does not have a significant effect on the DC conductivity of XLPE.

As shown in Figure 11, different stabilizer additions have a certain effect on the DC breakdown strength of XLPE composites. The breakdown strengths of the composites XLPE-A, XLPE-B and XLPE-C were 427.4, 450.5 and 405.4 kV/mm, respectively, which were significantly higher than the pure XLPE samples. It is known that the addition of a stabilizer has an effect of increasing the DC breakdown strength. When the electric field strength is large enough, electrons injected from the high voltage electrode get a lot of energy. The stabilizer inside the sample then captures electrons and reacts, causing the high-energy electrons to become low-energy electrons that hardly damage the XLPE molecular chain, ultimately increasing the breakdown strength of the composite. However, when the applied electric field on the

Figure 11. Relationship between the Weibull probability and the DC breakdown field of the different voltage-stabilizer modified XLPEs

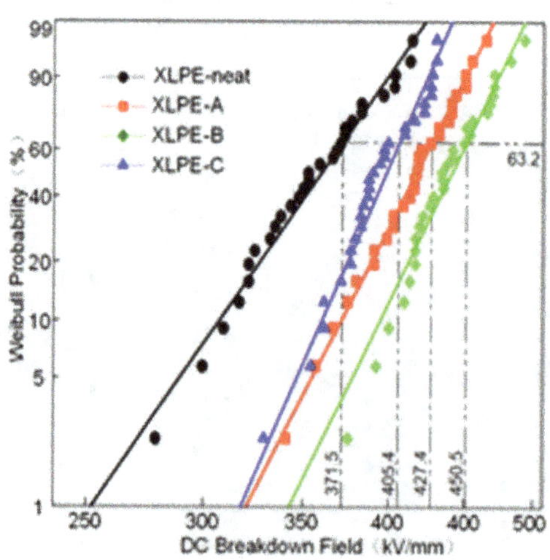

sample is low (for example, a low electric field applied in a conductivity experiment), the effect of the stabilizer is not significant. It is concluded that the higher the electron energy injected into the XLPE, the more effective the stabilizing effect of the stabilizer, which is very meaningful for improving the insulation stability in the practical application of the HVDC cable.

Space Charge Accumulation

Figure 12 shows that XLPE composites with different stabilizers have different dynamic space charge characteristics and electric field distributions at a DC electric field of -30 kV / mm. An enlarged view of the space charge distribution is intended to show the results more clearly. In pure XLPE, the same polarity charge injected from the anode is significantly increased. Moreover, the injected charge moves to the cathode as the polarization time increases, and a large amount of the sample, particularly the same polarity charge around the anode, is significantly aggregated. Eventually the electric field near the electrode is significantly distorted. In fact, the dynamic space charge characteristic in XLPE is the process of constant charge injection and extraction. When the DC electric field is large, the mobility of electrons is much larger than the mobility of holes, and the extraction of electrons causes positive space charges to accumulate around the anode.

Figures 12b, 12c and 12d show the space charge distribution of XLPE composites with different stabilizer additions. The results of the comparison test showed that the space charge inside the XLPE samples before and after the addition of A was basically the same, and the electric field distribution was similar, indicating that the stabilizer A has a weaker effect on suppressing the accumulation of space charge. For the samples XLPE-B and XLPE-C, only a small amount of space charge was injected into the XLPE sample, and most of the space charge was concentrated near the anode. That is, in most XLPE samples, the electric field is almost undistorted. The addition of stabilizers B and C is an extremely effective way to inhibit the space charge in the XLPE insulator.

As shown in Figure 13, the addition of different stabilizers also affects the space charge dissipation characteristics of XLPE under short-circuit conditions, which will help to analyze the effect of stabilizers on the space charge behavior of the composite. The positive space charge accumulates inside the sample during the polarization process, and most of the space charge in the material gradually dissipates as the depolarization time increases. It can

Figure 12. Space charge characteristics and electric field distortions in voltage-stabilizer modified XLPE samples as a function of polarization time under -30 kV/mm: (a) XLPE-neat; (b) XLPE-A; (c) XLPE-B; (d) XLPE-C

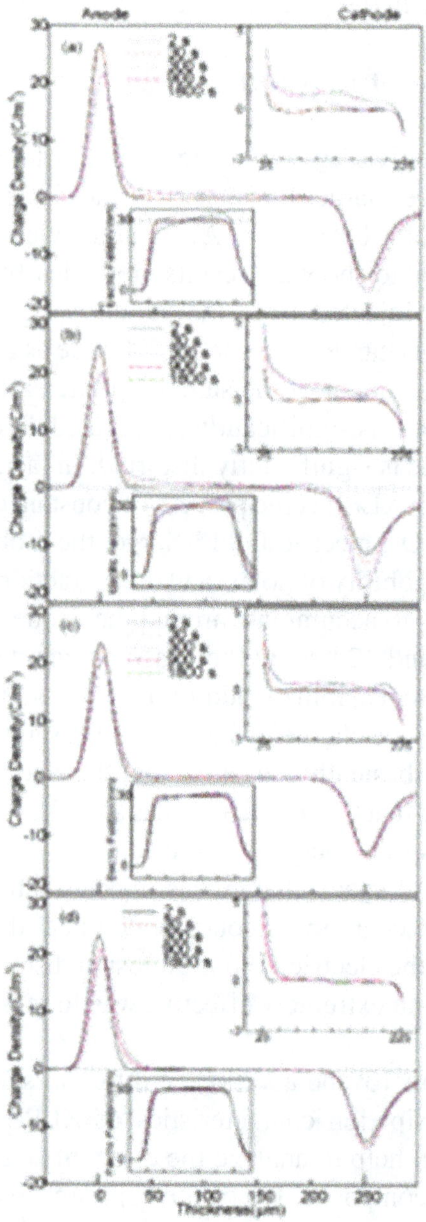

Figure 13. Space charge characteristics in voltage-stabilizer modified XLPEs as a function of depolarization time: (a) XLPE-neat; (b) XLPE-A; (c) XLPE-B; (d) XLPE-C

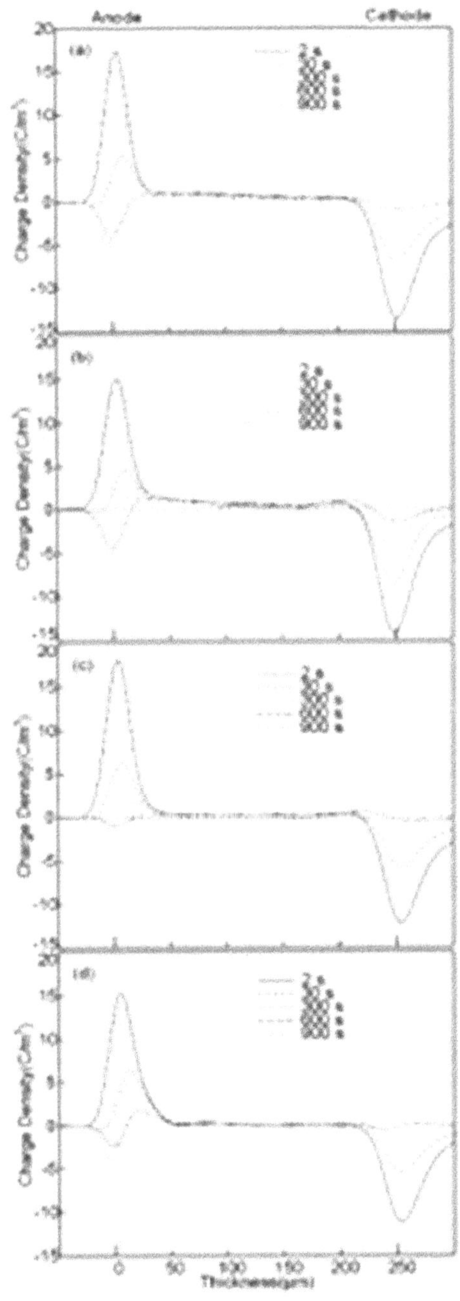

be seen that for pure XLPE samples, the space charge dissipation around the electrodes is very rapid but the space charge dissipation in the body is slower. The XLPE-B and XLPE-C samples have faster charge dissipation compared to the unadded XLPE, indicating that B and C have the ability to promote charge dissipation. After depolarization, the less space charge remaining inside the sample may be due to the reduced space charge injected internally during the polarization process. However, the XLPE-A sample did not exhibit the advantage of space charge dissipation characteristics.

Carrier Mobility and Trap Level Distribution

Space charge escapes from the trap during depolarization and is constantly dissipated, which is related to carrier mobility and trap level. Figure 14 shows the relationship between the average charge density value of the XLPE composite and the depolarization time after the addition of different stabilizers. The charge density of each sample is calculated by the following formula:

$$q(t)=\frac{1}{L}\int_0^L \left|q(x,t)\right| dx \tag{3-2}$$

Figure 14. Relationships between average charge density and depolarization of different voltage-stabilizer modified XLPEs

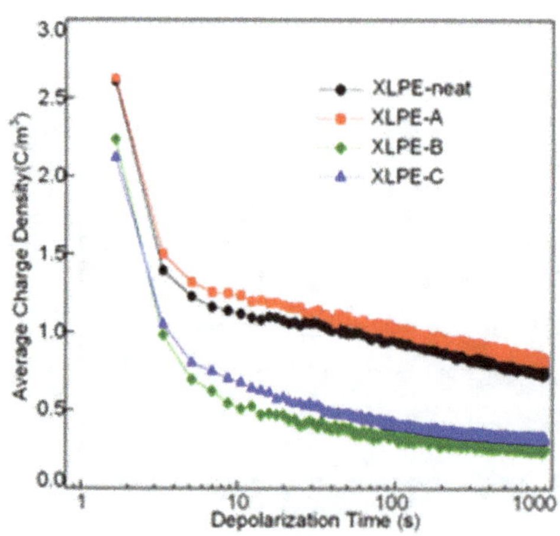

where, L is the thickness of the sample. $q(x, t)$ is the charge density in the specimen where x notes the location, and t notes the depolarization time.

From the test results, the average charge density in the pure XLPE sample was larger at 2 s. As the depolarization time increases, its charge density decreases at a non-linear rate. Since the charge release mechanism involves charges trapped by deeper traps, the decay rate of the charge density of the sample tends to decrease. Moreover, the initial charge density of XLPE-B and XLPE-C composites is significantly less than that of pure XLPE, while the density of residual charge after depolarization is much lower than that of pure XLPE.

As shown in Figure 15, the addition of stabilizers has a certain effect on the trap level distribution and carrier mobility of XLPE composite samples. As the depolarization time increases, the carrier mobility of all samples decreases. The apparent carrier mobility values of XLPE-B and XLPE-C are consistently higher compared to pure XLPE samples, especially during the first 10 s of the depolarization process. Moreover, the traps of XLPE-B and XLPE-C are also shallower.

Based on the above studies on space charge characteristics, carrier mobility and trap energy levels, the mechanism of action of stabilizers on space charge characteristics can be analyzed. The regulators B and C used in this experiment are suitable stabilizers for XLPE. The energy injected into the charge is higher, they can be absorbed and released in a harmless manner, until the last low energy charge remains. It is difficult for low-energy charge to enter most of the sample, space charge injection is reduced during polarization, and the charge accumulation inside the sample is also alleviated. At the same time, most of the low-energy charge is trapped by shallow traps after implantation, and the apparent carrier mobility of the XLPE composite containing the stabilizer is significantly higher during the depolarization process. It should be noted that the polarization process and the depolarization process are different. In the conductivity test, the carrier transport rate significantly affects the electrical conductance. During the polarization process, the conductivity of the XLPE composite is reduced due to the addition of a stabilizer. However, during depolarization, the carrier mobility depends on the rate of recapture of the trapped electrons. Stabilization causes electrons to be trapped in shallower traps, faster electron trapping and higher mobility of apparent trap control.

Figure 15. (a) Trap level distribution and (b) carrier mobility of different voltage-stabilizer modified XLPEs

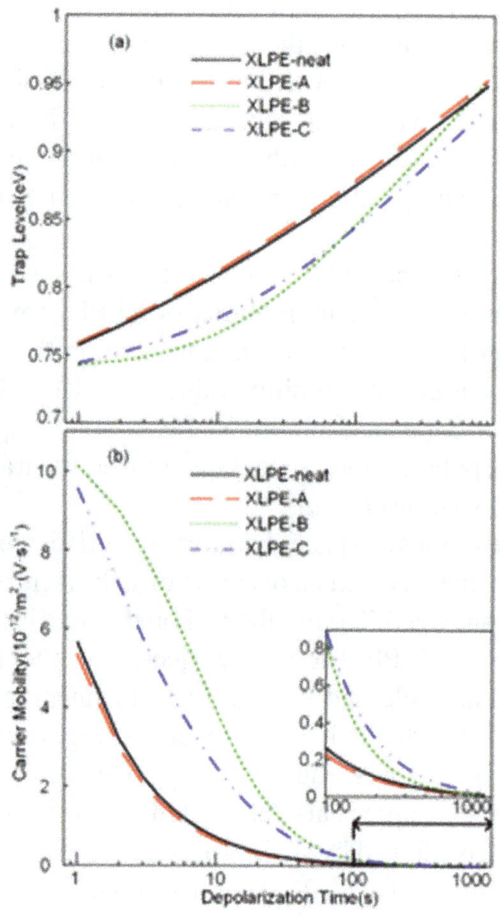

CONCLUSION

It can be seen from the research that GO has a significant improvement effect on the electrical properties and space charge characteristics of LDPE/GO nanocomposites. When the content of GO filler is 0.01 wt%, the relative dielectric constant and DC conductivity of nanocomposites are significantly lower than those of pure LDPE. However, excessive addition will result in a decrease in the electrical properties of the LDPE/GO composite. Therefore, a small amount of GO nanoparticles can ensure good dispersion of the nanoparticles and act to suppress space charge in the composite. The trap level density of LDPE will vary due to GO. At the same time, the electric field distortion factor confirms that GO can adjust the electric field distribution and reduce the electric field distortion.

Through thermal and mechanical experiments, it can be concluded that the compatibility of the stabilizer with the XLPE matrix is good. The DC conductivity test for XLPE materials indicates that the effect of the voltage stabilizer is small. However, stabilizers significantly affect the DC breakdown characteristics of XLPE. Regulators B and C have proven to be suitable choices to suppress space charge injection while promoting space charge dissipation, demonstrating a huge application prospect and important for XLPE insulation modification of HVDC cables.

REFERENCES

Bjorlow-Larsen, K. (2000). High voltage cables at the turn of the century. *IEEE Power Engineering Review, 20*(9).

Du, B. X., Han, C., Li, J., & Li, Z. (2019). Effect of Voltage Stabilizers on the Space Charge Behavior of XLPE for HVDC Cable Application. *IEEE Transactions on Dielectrics and Electrical Insulation, 26*(1), 34–42. doi:10.1109/TDEI.2018.007390

Du, B. X., Han, C., Li, Z., & Li, J. (2018). Effect of Graphene Oxide Particles on Space Charge Accumulation in LDPE/GO Nanocomposites. *IEEE Transactions on Dielectrics and Electrical Insulation, 25*(4), 1479–1486. doi:10.1109/TDEI.2018.006874

Du, B. X., Xu, H., Li, J., & Li, Z. (2016). Space charge behaviors of pp/poe/zno nanocomposites for hvdc cables. *IEEE Transactions on Dielectrics and Electrical Insulation, 23*(5), 3165–3174. doi:10.1109/TDEI.2016.7736882

Gray, A. P. (1970). Polymer crystallinity determinations by dsc. *Thermochimica Acta, 1*(6), 563–579. doi:10.1016/0040-6031(70)80008-9

Gross Hunt. (1967). *Dielectric Compositions Containing Halogenated Voltage Stabilizing Additives*. Patent US3350312.

Ieda, M. (1984). Electrical conduction and carrier traps in polymeric materials. *IEEE Transactions on Electrical Insulation, EI-19*(3), 162–178. doi:10.1109/TEI.1984.298741

Jarvid, M., Johansson, A., Bjuggren, J. M., Wutzel, H., Englund, V., Gubanski, S., ... Andersson, M. R. (2014). Tailored side-chain architecture of benzil voltage stabilizers for enhanced dielectric strength of cross-linked polyethylene. *Journal of Polymer Science. Part B, Polymer Physics*, *52*(16), 1047–1054. doi:10.1002/polb.23523

Jarvid, M., Johansson, A., Kroon, R., Bjuggren, J. M., Wutzel, H., Englund, V., ... Müller, C. (2015). A new application area for fullerenes: Voltage stabilizers for power cable insulation. *Advanced Materials*, *27*(5), 897–902. doi:10.1002/adma.201404306 PMID:25504254

Kim, Y. J., Ha, S. T., Lee, G. J., Nam, J. H., Ryu, I. H., Nam, S. H., ... Han, C. J. (2013). Investigation of space charge distribution of low-density polyethylene/go-gnf (graphene oxide from graphite nanofiber) nanocomposite for hvdc application. *Journal of Nanoscience and Nanotechnology*, *13*(5), 3464–3469. doi:10.1166/jnn.2013.7276 PMID:23858880

Lerf, A., He, H., Forster, M., & Klinowski, J. (1998). Structure of graphite oxide revisited||. *The Journal of Physical Chemistry B*, *102*(23), 4477–4482. doi:10.1021/jp9731821

Lewis, T. J. (1994). Nanometric dielectrics. *IEEE Transactions on Dielectrics and Electrical Insulation*, *1*(5), 812–825. doi:10.1109/94.326653

Lewis, T. J. (2002). Polyethylene under electrical stress. *IEEE Transactions on Dielectrics and Electrical Insulation*, *9*(5), 717–729. doi:10.1109/TDEI.2002.1038659

Li, Z., Du, B., Han, C., & Xu, H. (2017). Trap modulated charge carrier transport in polyethylene/graphene nanocomposites. *Scientific Reports*, *7*(1), 4015. doi:10.103841598-017-04196-5 PMID:28638056

Martinotto, L., Peruzzotti, F., & Del Brenna, M. (2001). *Cable, in Particular for Transport or Distribution of Electrical Energy and Insulating Composition.* Patent WO0108166.

Montanari, G. C., Laurent, C., Teyssedre, G., Campus, A., & Nilsson, U. H. (2005). From ldpe to xlpe: Investigating the change of electrical properties. part I. space charge, conduction and lifetime. *IEEE Transactions on Dielectrics and Electrical Insulation, 12*(3), 438–446. doi:10.1109/TDEI.2005.1453448

Montanari, G. C. (2011). Bringing an insulation to failure: the role of space charge. *IEEE Transactions on Dielectrics and Electrical Insulation, 18*(2).

Novoselov, K. S. (2004). Electric field effect in atomically thin carbon films. *Science, 306*(5696), 666-669.

Peng, S., He, J., Hu, J., Huang, X., & Jiang, P. (2015). Influence of functionalized MgO nanoparticles on electrical properties of polyethylene nanocomposites. *IEEE Transactions on Dielectrics and Electrical Insulation, 22*(3), 1512–1519. doi:10.1109/TDEI.2015.7116346

Pokharel, P., Lee, S. H., & Lee, D. S. (2015). Thermal, mechanical, and electrical properties of graphene nanoplatelet/graphene oxide/polyurethane hybrid nanocomposite. *Journal of Nanoscience and Nanotechnology, 15*(1), 211–214. doi:10.1166/jnn.2015.8353 PMID:26328332

Roy, M., Nelson, J. K., Maccrone, R. K., & Schadler, L. S. (2007). Candidate mechanisms controlling the electrical characteristics of silica/xlpe nanodielectrics. *Journal of Materials Science, 42*(11), 3789–3799. doi:10.100710853-006-0413-0

Singha, S., & Thomas, M. J. (2008). Dielectric properties of epoxy nanocomposites. *IEEE Transactions on Dielectrics and Electrical Insulation, 15*(1), 12–23. doi:10.1109/T-DEI.2008.4446732

Tanaka, T., Kozako, M., Fuse, N., & Ohki, Y. (2005). Proposal of a multi-core model for polymer nanocomposite dielectrics. *IEEE Transactions on Dielectrics and Electrical Insulation, 12*(4), 669–681. doi:10.1109/TDEI.2005.1511092

Wang, X., Yoshimura, N., Tanaka, Y., Murata, K., & Takada, T. (1999). Space charge characteristics in cross-linking polyethylene under electrical stress from dc to power frequency. *Journal of Physics. D, Applied Physics, 31*(16), 2057–2064. doi:10.1088/0022-3727/31/16/016

Wang, Y., Wang, C., Chen, W., & Xiao, K. (2016). Effect of stretching on electrical properties of ldpe/mgo nanocomposites. *IEEE Transactions on Dielectrics and Electrical Insulation*, *23*(3), 1713–1722. doi:10.1109/TDEI.2016.005733

Wutzel, H., Jarvid, M., Bjuggren, J. M., Johansson, A., Englund, V., Gubanski, S., & Andersson, M. R. (2015). Thioxanthone derivatives as stabilizers against electrical breakdown in cross-linked polyethylene for high voltage cable applications. *Polymer Degradation & Stability*, *112*, 63–69. doi:10.1016/j.polymdegradstab.2014.12.002

Zhang, H., Zhao, H., Wang, X., Shang, Y., Han, B., & Li, Z. (2014). Theoretical study on the mechanisms of polyethylene electrical breakdown strength increment by the addition of voltage stabilizers. *Journal of Molecular Modeling*, *20*(4), 2211–3281. doi:10.100700894-014-2211-y PMID:24699878

Chapter 7
Treeing Property In Polypropylene Under Various Temperature and Electrical Field

ABSTRACT

Polypropylene (PP) has no cross-linking process and environmentally friendly properties and is considered to be a replacement for cross-linked polyethylene (xlpe) for high voltage direct current (HVDC) cable insulation. High-voltage DC cable systems generate repetitive pulse voltages during operation and may encounter different temperature environmental challenges. This chapter discusses the effects of pulse amplitude and frequency on PP trees at different temperatures. A higher pulse frequency accelerates the propagation of the tree. Higher amplitudes accelerate tree growth and fractal dimensions. In addition, the effects of DC voltage, pulse voltage, and pulse frequency on the tree characteristics of PP at DC voltage and pulse combination voltage are also studied.

DOI: 10.4018/978-1-5225-8885-6.ch007

INTRODUCTION

HVDC transmission has been rapidly developed due to its large capacity and low loss. Therefore, there is an urgent need to develop high voltage DC plastic cables (Murata, 2006; Chen, 2015). During the operation of HVDC transmission cables, space charge accumulates, causing local electric field deformation of polymer insulators, accelerating the aging process, posing a great threat to safe operation (lan, 2014; Suzuki, 2013). Due to the decomposition of the crosslinking agent and the oxidizing agent, XLPE is defective as a high-voltage DC cable insulating material. These defects capture and accumulate more space costs (Han, 2011; Campus, 2002). Compared to crosslinked polyethylene, space charge is more difficult to accumulate in polypropylene due to the absence of cross-linking. Due to the non-crosslinking, recyclable environmental characteristics and excellent electrical and thermal properties of polypropylene, some researchers have suggested replacing polypropylene with polypropylene (Zhou, 2015; Holto, 2010).

Due to the on-off state and polarity reversal of the thyristor, the working pulse voltage is an overvoltage that often occurs in high-voltage DC converter transformers (Gao, 2013; He, 2013). The impact of the combination of surge voltage or DC voltage and surge voltage on the insulation of high voltage DC cables cannot be ignored. The high temperatures generated by the current flowing through the cable conductors will threaten the insulation operation of the cable. The design temperature of a power cable at rated load is typically 90°C, but typically up to 50-60°C (Clean Water, 2000). However, during a power outage, the temperature can reach 150°C in a short time (Bono, 1995). High temperature superconducting insulation is also facing the huge challenge of low temperature. At low temperatures, the mechanical properties of crosslinked polyethylene and polypropylene change. As the temperature decreases, the mechanical properties of the crosslinked polyethylene deteriorate. Therefore, it is important to study the resistance of polypropylene to repetitive pulse voltage and different temperatures.

The study found that voltage, temperature, frequency and other factors will affect the growth of trees. Chen et al. studied the effects of frequency and applied AC voltage on the characteristics of XLPE electrical trees. It is found that the higher the frequency, the faster the insulation strikes (Chen, 2011). The aging phenomenon of epoxy resin under positive and negative pulse voltage was studied. One is a simple straight tree - once a branch reaches the ground electrode and breaks. The other is a multi-branch tree -

when multiple branches reach the ground electrode, they fail. In this chapter, we studied the structure, wave characteristics, growth rate, fracture channel, cumulative damage and fractal dimension of trees, and studied the growth characteristics of trees at different temperatures. In addition, the effects of DC voltage, pulse voltage and pulse frequency on the tree are also studied. The chemical composition of the breakdown channel was analyzed by SEM, EDS and ATR-IR.

EFFECT OF LOW TEMPERATURE ON ELECTRICAL TREEING OF POLYPROPYLENE WITH REPETITIVE PULSE VOLTAGE

Tree Growth Characteristics

Figure 1 shows a typical dendritic structure of polypropylene and crosslinked polyethylene tested at different temperatures. The pulse amplitude and frequency are 12 kV and 400 Hz, respectively (Du et al., 2016). The results show that the tree structure of cross-linked polyethylene changes with temperature, and the dendritic structure of polypropylene changes with temperature. In xlpe, the temperature of the shrub tree structure is 30°C, and the temperature of the branched tree structure is -30° and -196°c. The stagnation tree and the branching pine were observed at -90°c. At -90°C, the twig pine is a double tree, part of which is a branch structure and the other part is a pine structure. The other tree observed at -90°C is a stagnant tree that stagnates when the tree grows to a certain length. In polypropylene, the structure of the branches is 30°, -30°, -90°, and -196°C. The density of the branches is different at different temperatures. At -90°C, the shoots are denser than -30°C, the side branches produced by the main branches are much longer than -30°C, and the number of main branches is less than -30° and -90°C at -196°C. The main branches are spaced apart. At the same temperature, the cross-linked polyethylene has a denser tree structure, and the main branches produce a large number of small branches.

The length of this tree is the longest branch in the direction of the electric field (Figure 2). Fractal dimensions measure the complexity of a branch. Therefore, using the box counting method, the value of the fractal dimension varies between 1 and 2, which is consistent with the previous literature.

Figure 1. Typical profiles of electrical trees in the XLPE and the PP under different temperatures with the pulse voltage of 12 kV and 400 Hz

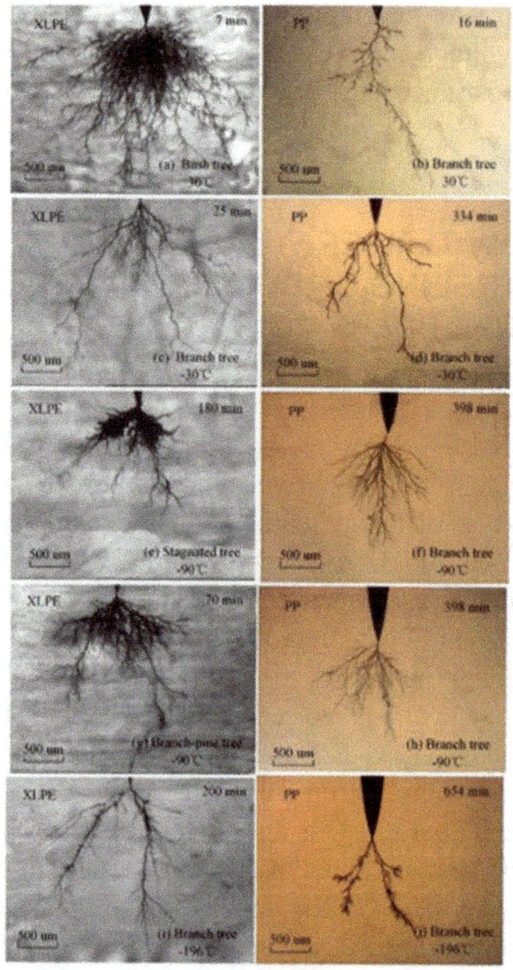

At 30°C, when the length of the crosslinked polyethylene reaches 2000 μm in 7 minutes, the length of the tree of polypropylene is only 1120 μm; when the planting time reaches 16 minutes, the length of the tree of polypropylene is 1907 μm; in the crosslinked polyethylene Before the decomposition at 30°C, the fractal dimension of the crosslinked polyethylene was 1.767, while the fractal dimension of the polypropylene was only 1.392, and then slowly increased. The branching pine in cross-linked polyethylene is selected as a typical tree

Figure 2. The propagation characteristics of electrical trees in the PP and the XLPE at 30, -30, -90 and -196 °C

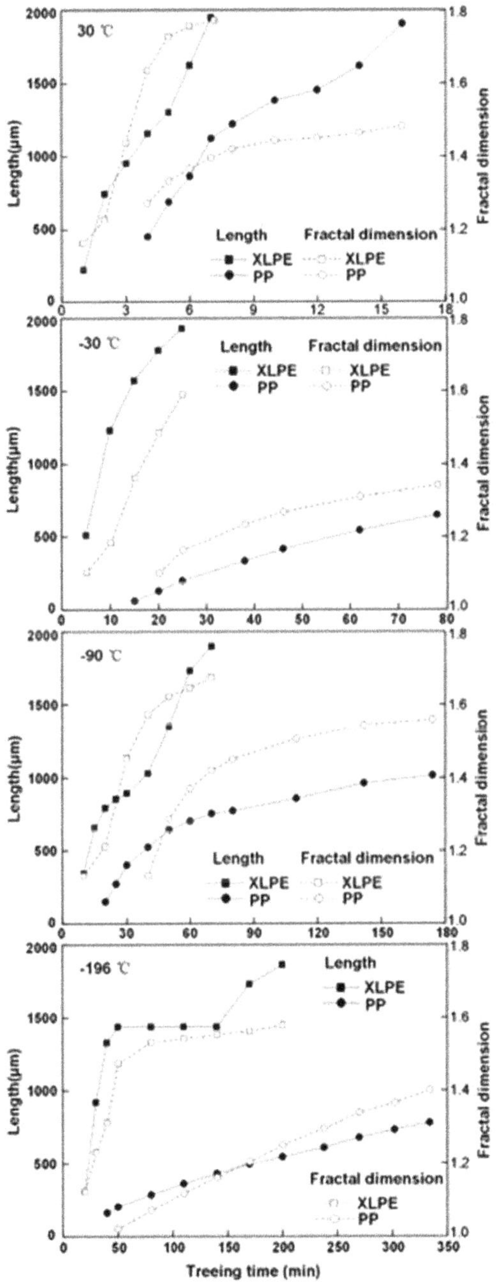

compared with the branched tree in polypropylene. The decomposition of branched pine in cross-linked polyethylene occurs. At -196°C, when the length of the crosslinked polyethylene reached 2000 μm in 220 minutes, the tree length of pp was only 570 μm and then slowly increased. When the planting time reaches 334min, the tree length of pp is only 779μm. At lower temperatures, the growth rate of electric branches is significantly reduced. The charge will be trapped by these defects and accumulate in the body, resulting in more partial discharge. Therefore, the molecular chain in the crosslinked polyethylene is more likely to break. This is why the tree length and fractal dimension of cross-linked polyethylene are definitely greater than the tree length and fractal dimension of polypropylene at the same age.

In order to study the difference in growth of polypropylene trees and cross-linked polyethylene trees at low temperatures, the initial probability of the electron tree was calculated. The percentage of 20 samples of the 30 μm original tree was generated in 10 minutes. As shown in Figure 3, when the temperature is varied between -30° and -196°C, the probability of tree start-up is significantly reduced. Take the probability of branching in polypropylene as an example, only 5% and 65% at -196° and -90°C, respectively, and up to 90% at -30°C. As shown in Figure 3, the lower temperature tip is difficult. Produce electric branches. As can be seen from Figure 3, the probability of cross-linked polyethylene at -30°C is 95%, while the probability in polypropylene is 90%. The crosslinked polyethylene is 85%, the polypropylene is 65%, and the temperature is -90°C. At -196°C, the crosslinked polyethylene was 35% and

Figure 3. The tree inception probability for 10 minutes in the PP and the XLPE under low temperature

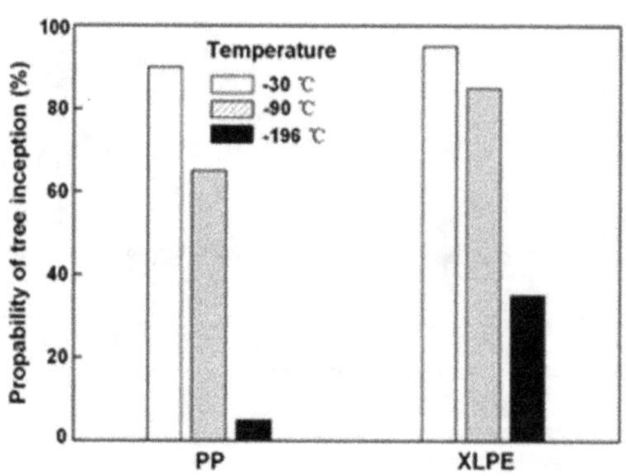

the polypropylene was 5%. The results show that the starting probability of electron tree in cross-linked polyethylene is higher than that of polypropylene, indicating that the possibility of tip-forming trees in cross-linked polyethylene is greater than that of polypropylene.

Effects of Pulse Frequency

The initiation and reproduction of trees depend on Pd, oxidation and chemical decomposition activities in the micro-branched channels. The structure of the tree may be completely different as the ambient temperature and pulse frequency change. Figure 4 shows the dendritic structure of polypropylene at different pulse frequencies at low temperatures. Pre-decompose the selected tree image to get a complete tree structure image.

As shown in Figure 4, the structure of the tree changes with temperature. Branched trees differ from -30°C at -196°C. At -196°C, the gap between the main branches becomes larger, and more small branches appear between the main branches. In addition, at lower temperatures, the passage of trees becomes darker and rougher. At -90°C, the shrubs are dense structures. The higher the frequency, the more the number of main branches and side branches, the more overlapping the tree channels, resulting in higher tree density.

When a pulse voltage is applied, charge is injected into the insulating material from the tip of the needle, and the energy released by the trapping

Figure 4. Typical profiles of electrical trees with different pulse frequencies under different temperatures with the pulse voltage of 15 kV

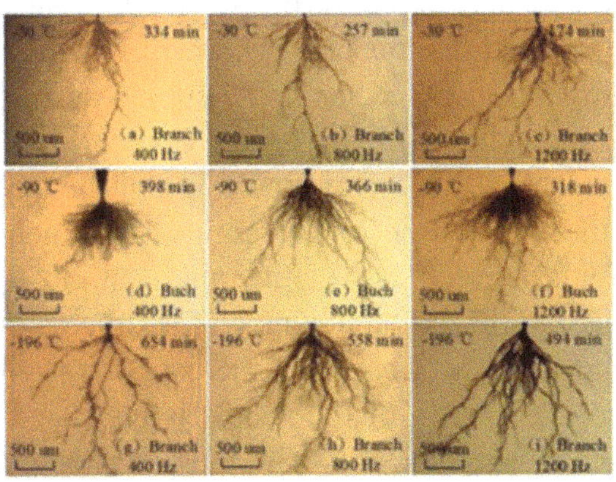

and releasing process of these charges will destroy the molecular chain around the tip of the needle, thereby forming a micro-tree-like channel. In addition, pd will release more charge to the tree channel, releasing local heat and high pressure, further destroying the molecular chain. The higher the frequency, the stronger the pd activity and the greater the amount of space charge generated by pd. In addition, more charge is injected per unit time, thereby colliding with the molecular chain.

At low temperatures, partial discharge will be greatly diminished. First, the length of the tree grows at a high speed. Then the main branches grow slowly, and the main branches create new channels and cross each other. In the last period, the tree grew very fast. The conductivity of trees can be inferred from the color of the tree channel. At -30, -90, and -196°C, the conductivity of the trees is higher because the black tree channel means residual heavy carbon in the tree channel and can be used as an equivalent needle electrode. According to the equation of maximum electric field strength at the reference tip, as the distance between the tip and the ground electrode becomes shorter, the electric field will be higher. At 800 and 1200 Hz, the growth trend of -90°C is related to the growth trend of -30°C. However, when the frequency of 400 Hz is used at -90°C, the length of the tree initially increases rapidly, then almost stops increasing, and finally increases again. During this period, the length of the trees almost stopped growing due to the formation of shrubs. When the frequency is high, charge and discharge and partial discharge are sufficient to form a shrub area while promoting the growth of the main branches. This is why in the second part, the length of the tree increases slowly at 800 and 1200 Hz, and almost stops increasing at 400 Hz. - The growth trend at 196°C at different frequencies is related to the growth trend at -90°C at 400 hz. As the temperature decreases, the partial discharge decreases, and the heat and pressure during the discharge process are low, which is insufficient to promote tree growth. In the partial discharge process, when local high temperature and high pressure accumulate, the electric field increases as the distance between the tree tip and the grounding pole decreases. These two factors promote the rapid growth of trees at -196°C. The conclusion of the tree growth trend is that the length of the tree first grows at a high speed. Then the main branch grows slowly. In the last period, trees grew rapidly at -30°, -90° and -196°C.

As shown in Figure 5, at the same temperature, the length increases with increasing frequency. As the frequency increases from 400hz to 1200hz, the charge injected into the tip of the needle is increased more and more, and the partial discharge activity is enhanced. In this case, as the frequency increases, the length of the electric tree tends to be longer. As shown in Figure

Figure 5. The propagation characteristics of electrical trees in the PP with the pulse frequency of 400, 800 and 1200 Hz at -30, -90 and -196 °C

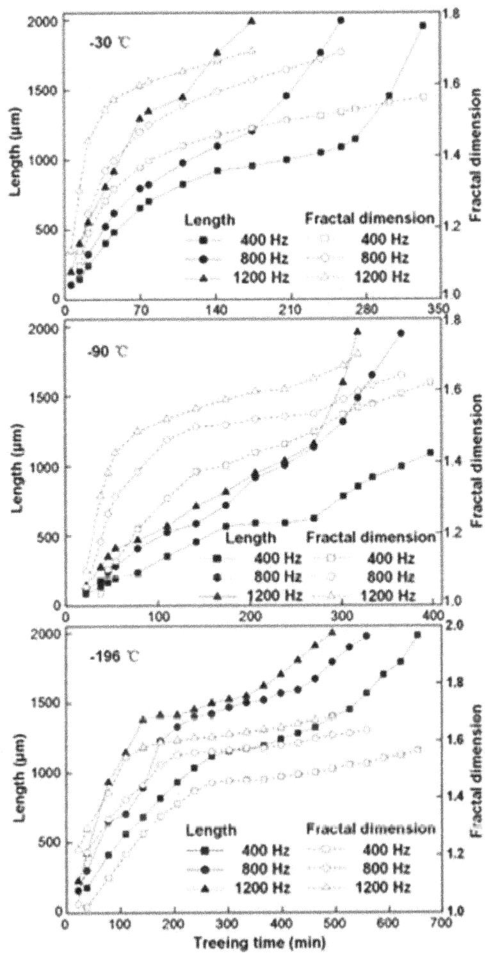

5, the fractal dimension changes in each case are similar. In the initial stage, strong pd destroys the molecular chain, causing the main branch to be rapidly initialized to form a branched structure. During the breeding process, although the growth of the tree grows slowly or almost stops growing, it will produce more branches, resulting in a slow increase in fractal dimension. It is worth noting that the growth rate of fractal dimension at -196°C is much slower than that at -30° and -90°C, as pd decreases with decreasing temperature and between -168°C The gap is large and it is difficult to produce branches and cross each other. In the final stage, some small channels grow rapidly onto the ground electrode, causing the fractal dimension to increase slowly. As can be seen from Figure 5, the fractal dimension increases with increasing frequency. As the frequency increases, a large amount of charge may collide with the polymer, thereby destroying the molecular chain. Therefore, a new degradation zone is formed. The main branches produce more side branches that may overlap each other during the expansion process. This causes the branches to pile up and the complexity of the electrical branches to increase accordingly.

2.3 Effects of Pulse Amplitude

The pulse amplitude is 12 and 15 kv and the frequency is 400 hz. At -30°C and -196°c, the structure of the tree is branched, but the branching density increases with increasing amplitude. Furthermore, at -30 and -196°C, the length of the 15 kv amplitude tree is longer than the length of the 12 kv amplitude tree. At -90°C, the structure of the tree changed from a branch to a shrub, with a large variation. This branch develops mainly along the direction of the 12 kV electric field. Even perpendicular to the direction of the electric field, the amplitude is 15 kv, and more of the side branches extend along a wider arc angle around the tip of the needle. The larger the amplitude, the stronger the charge effect, resulting in more partial discharge. The higher the pressure and heat generated in the tree channel, the more the distribution of molecular chains in the random direction will be destroyed, so the tree channel is easier to be in the wider arc angle.

As shown in Figure 7, the growth rate of the tree decreases with decreasing temperature because the partial discharge decreases with decreasing temperature. In addition, the growth mode of the electric tree is different from the growth mode of temperature and amplitude. At -30°C, the growth trend of tree length is the same at different pulse amplitudes. First, the length of the tree increases rapidly. Then the main branch grows slowly. In the last period, trees grew faster than before. At an amplitude of 15 kv at -90°C, the length of the tree initially increases rapidly, then almost stops increasing, and finally increases again. However, at -90°C, when the amplitude is 12 kv, the growth trend increases rapidly in the initial stage, and then continues to increase at a lower rate than the initial stage over a period of time. The growth trend of -196°C and 15 kv is related to the growth trend of -30°C, while the growth

Figure 6. Typical profiles of electrical trees with different pulse amplitudes under different temperatures with the pulse frequency of 400 Hz

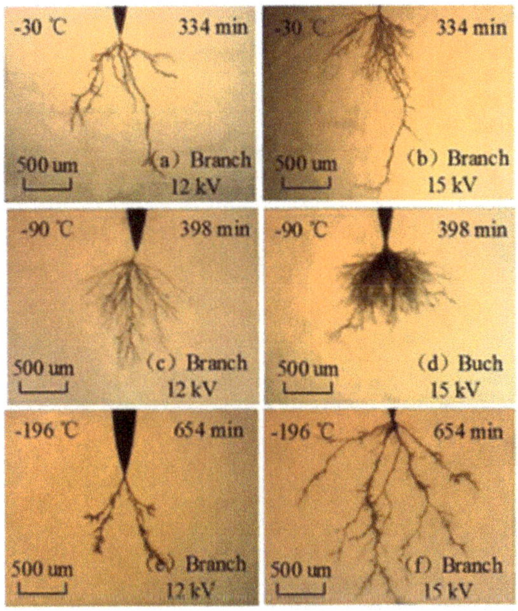

trend of -196°C and 12 kv is different. When the trees reached 920 μm in length, they almost stopped growing.

At different temperatures, the trend of fractal dimension is similar. It grew rapidly in the early stages and then continued to grow at a lower rate than initially.

Tree Structure Distribution

At -30°C, the structure of the branched tree is related to the appropriate heat and pressure generated by the 15 kv 400 hz stress. As the temperature decreases, the hardness and modulus of elasticity of the polypropylene increase, which means that the polypropylene at -90°C is more fragile than the polypropylene at -30°C and is more susceptible to cracking. Due to charge injection or ultraviolet degradation, the molecular chain is broken to form micropores. Among them, electronic avalanche and partial discharge occur. If the charge generated by the avalanche discharge is injected into an existing branch or a local microcrack generated by a mechanical crack, a denser region is formed. Therefore, dense regions are more easily formed at -90°C than at

Table 1. Typical structure of electrical trees at different conditions

Amplitude	Frequency	-30 °C	-90 °C	-196 °C
12 kV	400Hz	Branch	Branch	Branch
15 kV	400Hz	Branch	Bush	Branch
15 kV	800 Hz	Branch	Bush	Branch
15 kV	1200 Hz	Branch	Bush	Branch

Figure 7. The propagation characteristics of electrical trees in the PP with the pulse voltage of 12 and 15 kV at -30, -90 and -196 °C

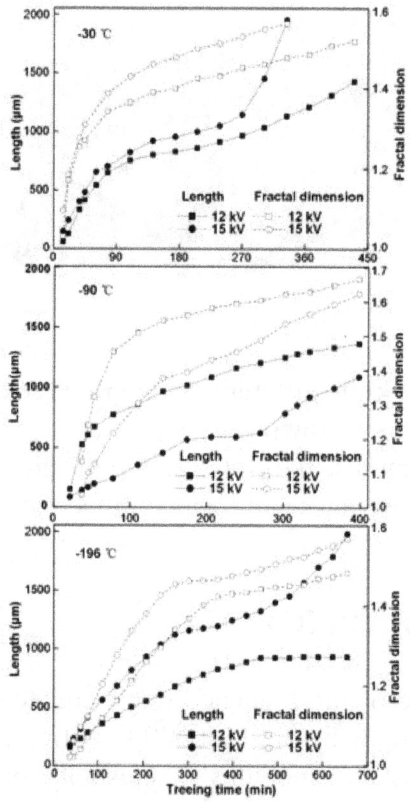

-30°C. Thus, at a temperature of 15 kV from -30° to -90°C, the structure of the tree changes from a branch to a shrub. When a radical reacts with oxygen, there is an auto-oxidation process that produces hydrogen peroxide. However, unstable hydroperoxides may decompose and regenerate free radicals. In the above process, the polymer chain is broken. Therefore, the structure branches at -196°C, and when the experimental pulse voltage is increased from 12 kV to 15 kV at -90°C, partial discharge is more likely to destroy the molecular chain of pp, thereby forming a denser region.

EFFECT OF AMBIENT TEMPERATURE ON ELECTRICAL TREEING AND BREAKDOWN PHENOMENON OF POLYPROPYLENE WITH REPETITIVE PULSE VOLTAGE

Tree Structure and Tree Initiation Characteristics

Figure 8 shows the typical shape of a tree under different conditions (Du et al., 2016). The selected tree image is the image before the decomposition, and the entire tree structure image is obtained. The tree time of the picture in Figure 8 is different. The structure of the electronic tree is dendritic and

Figure 8. Typical morphologies of electrical trees under different conditions

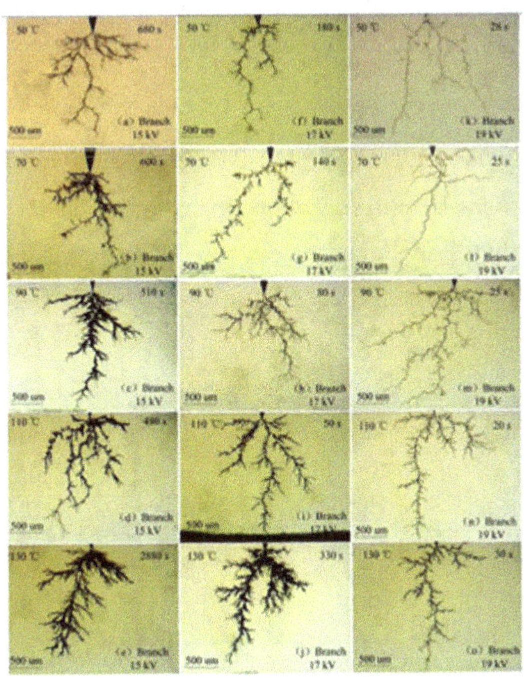

does not change with changes in temperature and pulse amplitude, but its density varies with temperature. As the temperature increases, the density of the shoots increases significantly and the branches become deeper and deeper. The results show that temperature has an effect on the structure of the trees. At the same temperature, as the amplitude increases, more side branches around the needle electrode extend in all directions.

Figure 9 shows the relationship between the probability of starting different pulse amplitudes and the 5 s temperature. Calculate the probability of activation when the pulse voltage is applied to the sample for 5 s. As the temperature increases from 50°C to 130°C, the startup probability at 15 kV changes from 5% to 45%. The initial probability varies from 30% to 75% at 17 kV. The initial probability varies from 55% to 95% in the range of 19 kV. The results show that the probability of starting is related to temperature.

The molecular chain produces free radicals. Macromolecular radicals initiate chain reactions and form new traps. As a result, the probability of initiation of the tree increases as the temperature increases from 50°C to 130°C. The tree's probability of activation increases as the amplitude increases from 15 kv to 19 kv. It is worth noting that at 15 kV, the tree has a probability of starting at 45%, while at 130°C, at 17 kV and 19 kV, the tree has a probability of starting at 75% and 95%, respectively. The larger the pulse amplitude, the greater the energy injected into the charge, causing severe damage to the tip.

Figure 9. Relation between the inception probability and the temperature with different pulse amplitudes at 5s

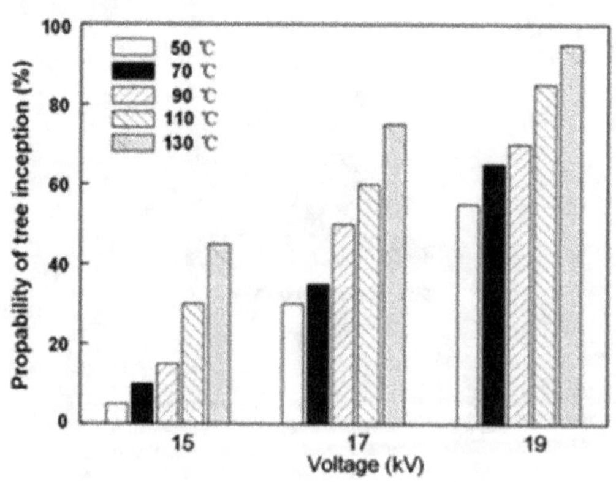

Treeing Characteristics

In Figure 10, the tree time ratios for different pulse amplitudes are different. The pulse amplitudes are 15, 17 and 19 kv, respectively. The growth rate increases as the temperature increases from 50°C to 110°C, and the growth rate at 130°C is lower than 110°C.

The dsc curve is shown in Figure 11. During the cooling process, an exothermic peak in the heat flow temperature curve was observed, corresponding to the crystallization of the polymer chain at 122°C; during the heating process, an endothermic peak in the heat flow temperature curve was observed, corresponding to the microcrystal at 148°C The melting point.

As shown in Figure 11, as the temperature increases from 110°C to 130°C, the crystal begins to melt. It is easier for electrons to accelerate the formation of more "hot electrons" in the free volume at 130°C. These hot electrons collide with the molecular chain to produce more A dendritic channel that forms a denser region at 130°C. At the same time, due to the melting of the crystal, the elastic modulus is lowered at 130°C. Dense area.

Figure 12 shows the relationship between fractal dimension and tree time for different pulse amplitudes. The fractal dimension is estimated using the box number method. At the same time, with the change of amplitude from 15kv to 19kv, the fractal dimension shows an increasing trend. As the pulse voltage increases from 15kv to 19kv, the fractal dimension changes from 1.375 to 1.586 at 90°C and 25s. At 110°C and 20s, the fractal dimension changes from 1.423 to 1.604, indicating that the larger the pulse amplitude, the more the fractal dimension. At the same temperature, high amplitude will accelerate the growth of trees. The greater the amplitude, the stronger the charge action and pd activity, helping to form longer branches and denser branches..

Breakdown Phenomenon

Figure 13 shows the relationship between the fractal dimension of the breakdown tree and the breakdown time (TTB) under different conditions. If a malfunction occurs, turn off the power immediately. Then use the camera to take an image of the electronic tree and calculate the fractal dimension and cumulative damage. These results are the fractal dimension and cumulative damage of the fault tree. At the same temperature, as the pulse amplitude increases, ttb tends to be shorter because high pulse amplitudes accelerate tree growth. The dimension increased from 50° to 110°C; at 19 kV, the fractal

Figure 10. Relation between the tree length and the treeing time under different temperatures

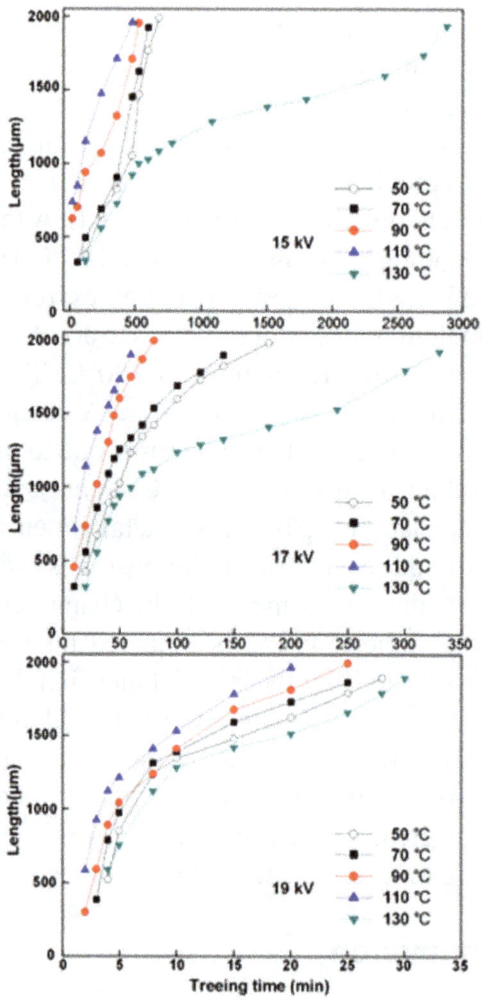

dimension increased from 1.803 to 1.827; when 50° increased to 110°C, the fractal dimension increased from 1.811 to 1.811. At 130°C, it can be seen from Figure 13 that the fractal dimension of the decomposition tree increases with the increase of temperature. In addition to the different amplitudes at 130°C, the thermal motion of the molecular chain increases with the increase

Figure 11. DSC analysis

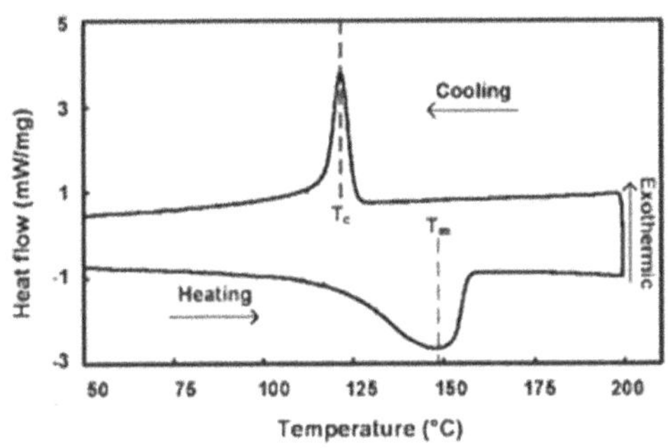

of temperature. Become active. The molecular chains around the tree wall are more susceptible to breakage. Although the electric field at the tree tip at 130°C is greater than the electric field at 110°C, it helps to reduce the discharge at 130°C. Therefore, the fractal dimension of the breakdown tree at 130°C is greater than 110°C. It is smaller than the temperature at 110°C.

At 15 kV, cumulative damage decreases as the temperature increases from 50° to 90°C. Less than 110°C at 90°C and less than 110°C at 130°C. Cumulative damage at 70°C is greater than 90°C at 17 kV. Less than 110°C at 90°C and less than 110°C at 130°C.

The cumulative damage that breaks through the tree refers to the number of pixels in the area covered by the breakdown of the tree, which is used to indicate the damaged area of the insulating material. Figure 14 shows the relationship between cumulative damage and fault tree TTB under different conditions. At 15 kV, the cumulative damage decreases as the temperature increases from 50 degrees Celsius to 90 degrees Celsius. Less than 110 degrees Celsius at 90 degrees Celsius and less than 110 degrees Celsius at 130 degrees Celsius. At 70 degrees Celsius, the cumulative damage is greater than 90 degrees Celsius at 17 kV. It is less than 110°C at 90°C and less than 110°C at 130°C. At 19 kV, the cumulative damage was highest at 50 °C and decreased with increasing temperature to 130 °C, indicating that the cumulative damage to the tree was not associated with high temperatures.

Figure 16a shows the surface topography and edax results of the original sample. Figures 16b and 10c show the surface topography and edax results for a typical breakdown channel. The voltage applied in Figure 16b has an amplitude of 17 kV and a temperature of 70°C. The voltage applied in

Figure 12. Relation between the fractal dimension and the treeing time with different pulse amplitudes

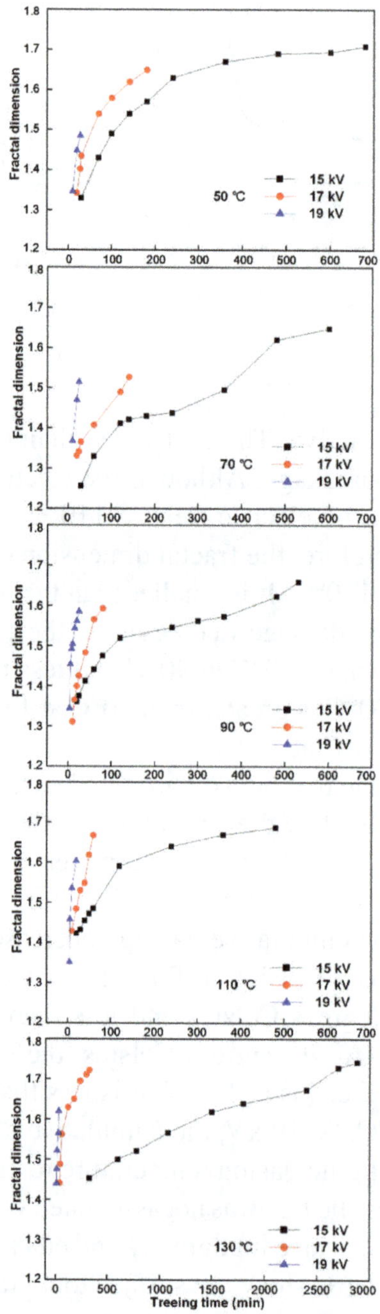

Figure 13. Relation between the fractal dimension and the TTB of breakdown electrical tree under different conditions

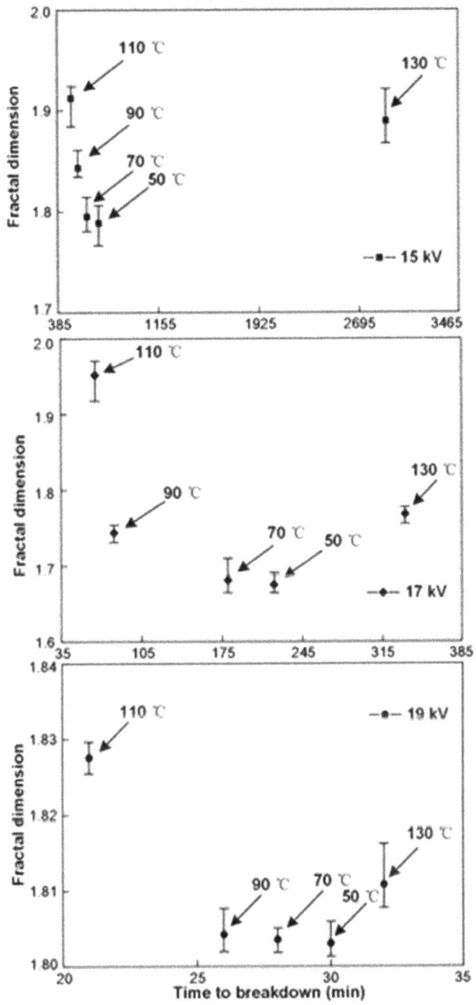

Figure 3.3c has an amplitude of 17 kV and a temperature of 130°C. Figures 10b and 10c show different particle morphologies. As shown in Figure 16b, the observed particle morphology is agglomerated, while in Figure 16c, the observed particle morphology is dispersed and significant ablation cracks

Figure 14. Relation between the accumulated damage and the TTB of breakdown electrical tree under different conditions

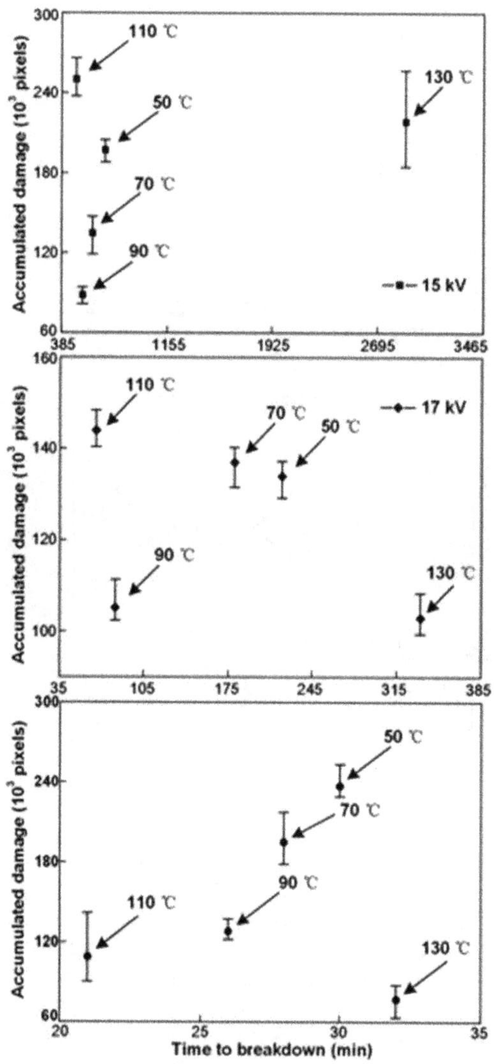

are found. Note that due to the limitations of edax analysis, it is not possible to find hydrogen atoms in the sample. As shown in Figures 16a, 16b and 16c, carbon and gold atoms were observed in the original sample and in the

Figure 15 Typical breakdown morphologies of electrical trees under different conditions

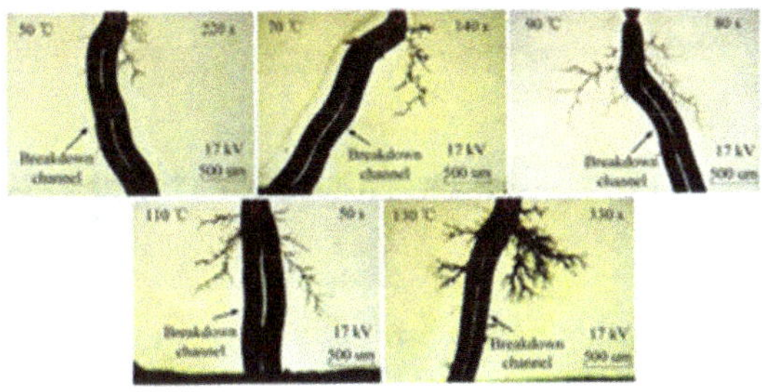

decomposition channel. The gold atoms on the surface of the sample originate from the metallization process before scanning electron microscopy and are considered to be impurities, so the content of gold atoms is not considered. Oxygen is introduced into the decomposition channel as compared to the analysis of the original sample. In Figure 16b, oxygen accounts for 10.32% by weight and atoms account for 7.95%. In Figure 16c, oxygen accounts for 16.03% of the weight and atoms account for 12.53%. The atr-ir spectrum of the breakdown channel has a new absorption peak at 1730 and 1253 cm-1 compared to the atr-ir spectrum of the original sample. In an anaerobic environment, the injected charge causes the molecular chain to break, forming a small cavity and an electric tree after a long period of application of the pulse voltage. Oxygen plays an important role in the growth of trees in an oxygen-rich environment. When the free radical reacts with oxygen, the oxidation process begins. Free radicals react with oxygen to form hydroperoxides. Unstable hydroperoxides decompose and regenerate free radicals.

$RH \rightarrow R^* + H^*$; $R^* + O_2 \rightarrow ROO^*$;

$ROO^* + RH \rightarrow ROOH + R^*$; $ROOH \rightarrow RO^* + OH$

Similarly, in this process, C = O and C-O groups were formed, which were observed by ATR-IR spectrum.

Figure 16. SEM observation and EDAX analysis of the original sample and breakdown channels

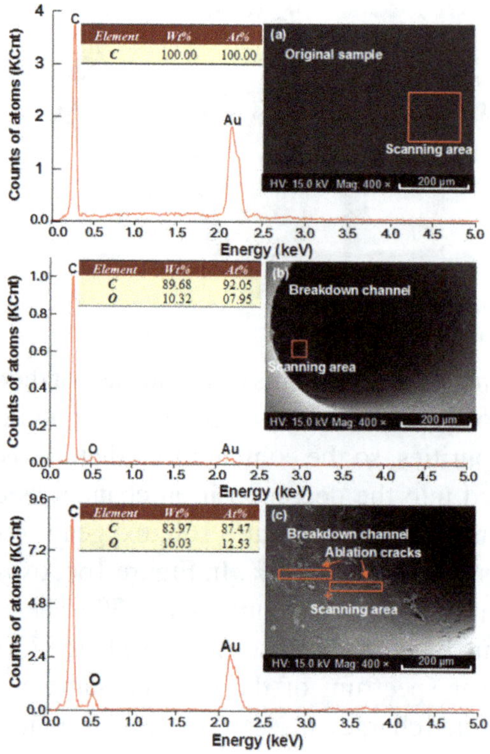

Table 2. Absorption peaks of common group

Wave Number (cm^{-1})	Group
3000-2850	C-H stretching vibration
3100-3010	C-H stretching vibration
1465-1340	C-H bending vibration
1000-675	C-H out of plane bending vibration
880-680	C-H out of plane bending vibration
1300-1000	C-O stretching vibration
1750-1700	C=O stretching vibration
1720-1706	C=O stretching vibration

Figure 17. ATR-IR spectra of the (a) original sample and (b) breakdown channel

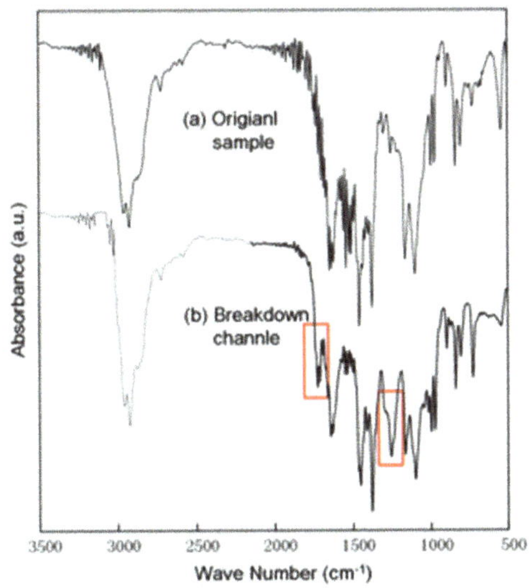

ELECTRICAL TREEING INITIATION AND BREAKDOWN PHENOMENON IN POLYPROPYLENE UNDER DC AND PULSE COMBINED VOLTAGES

Effects of Pulse Voltage Combined With the Different DC Voltages

Figure 18 shows the waveform of the combined DC voltage and pulse voltage

Figures 18A1, 18A2, and 18A3 represent a combination of +25 kV pulses and +5, +15, and +25 kV DC voltages, respectively. Figures 18b1, 18b2, and 18b3 represent a combination of a +25 kV pulse voltage and a -5, -15, and -25 kV DC voltage, respectively. The pulse frequency is 400 Hz (Du et al., 2019).

Figure 19 shows a typical tree structure with different DC amplitudes and a surge voltage of +25 kV. Planting time is 25 minutes. At positive DC voltages, as shown in Figures 19a1, 19a2 and 19a3, the tree structures have similar amplitudes but different amplitudes, which means that the positive DC voltage has little effect on the tree structure.

Figure 18. Waveforms of DC and pulse combined voltages

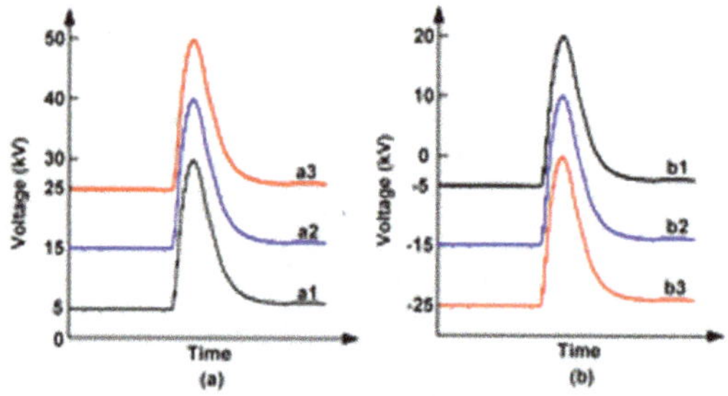

Figure 19. Typical electrical tree structures of different DC amplitudes combined with +25 kV pulse voltages

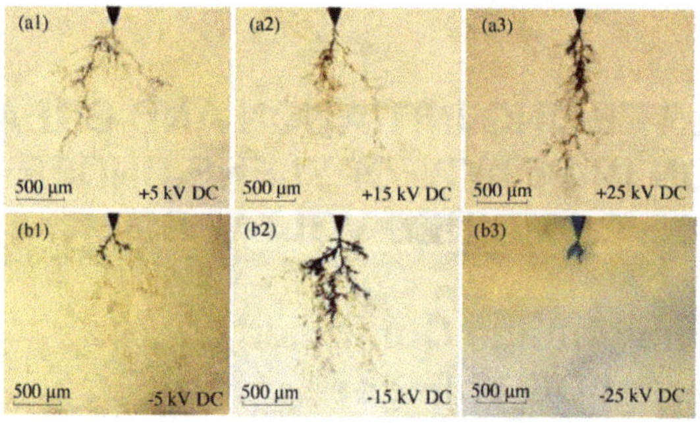

When a positive DC voltage is applied, charge is injected from the tip of the needle. These uniform space charges are formed around the tip or the electrical tree, reducing the electric field. This is also the reason why the startup voltage of the electric tree is high under DC voltage. When a pulse voltage is applied, a large amount of charge is instantaneously injected; these moving charges stimulate the existing charge, causing it to break from the tip of the needle or electrical branch and re-enter other traps. Due to the trapping process, the charge state transitions from a high energy state to a low energy state. During this process, the hot electrons destroy the molecular chain and form a low density region. Avalanche discharges also occur, causing molecular chain breaks. The larger the DC voltage amplitude, the more charge is injected.

Figure 20. Relationship between the tree length and the treeing time of different DC amplitudes and +25 kV pulse combined voltages

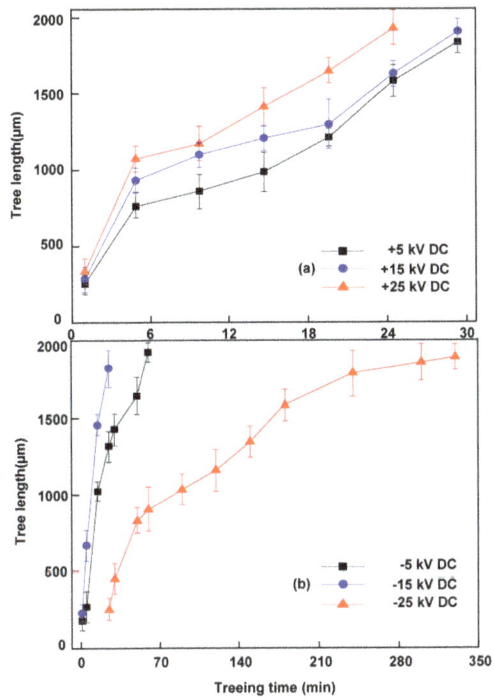

Therefore, the growth rate increases as the DC amplitude increases. As shown in Figure 20b, as the DC voltage increases from -5 dc to -15 kV, the growth rate increases. Figure 21b shows the charge transfer process for -5 and -15 kV, respectively, corresponding to Figures 18b1 and 18b2. A simulation model based on the charge transfer process was established. The peak electric field strength of -15kv DC +25kv pulse is large, indicating that the polarity change effect is larger at -15kv. However, as the DC voltage increases from -15 kV to -25 kV, the growth rate decreases. As shown in Figure 21c, the charge transfer process at -25kv DC is different from the charge transfer process at -5 and -15kv DC, corresponding to the voltage applied in Figure 18b3. When a pulse voltage is applied, electrons are extracted due to a decrease in the potential of the tip. This charge motion causes molecular chain breaks and electron tree growth. Due to the lack of polarity change process, the growth rate of -25kv DC is less than the growth rate of -5 and -15kv DC.

Figure 21. Charge transport processes under different DC amplitudes

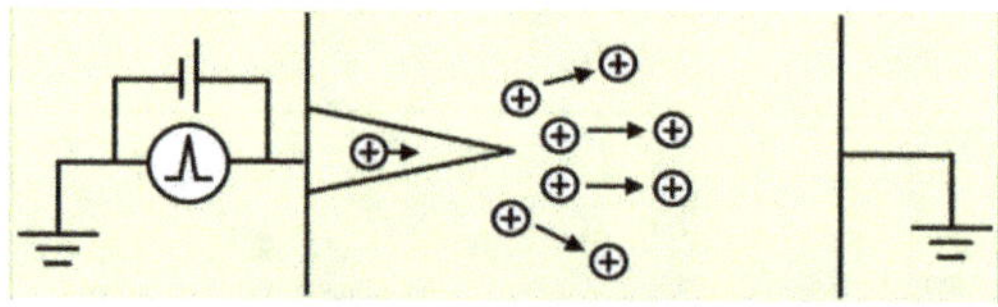

(a) Positive DC and pulse combined voltages.

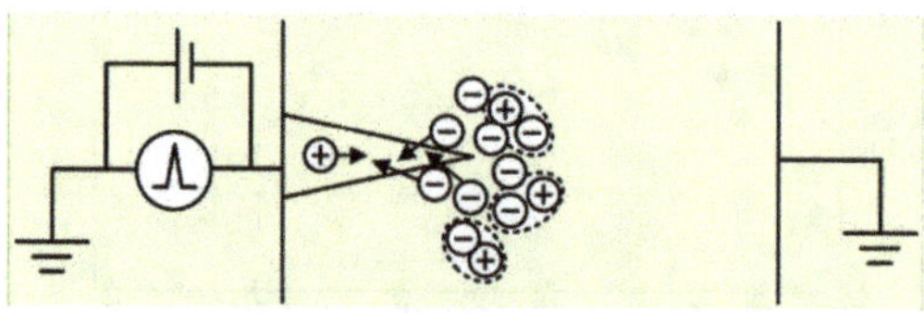

(b) -5 and -15 kV DC and pulse combined voltages.

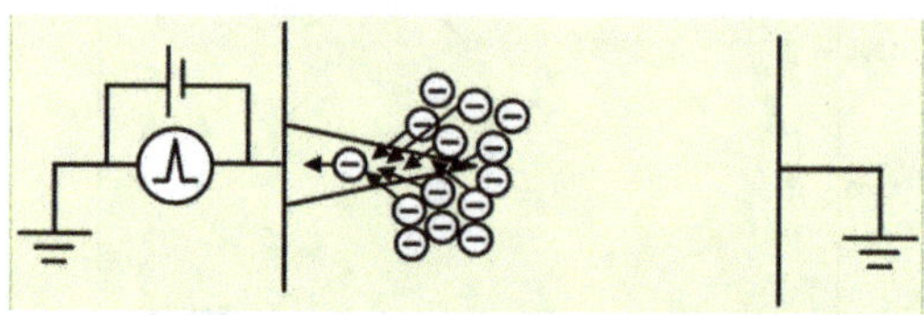

(c) -25 kV DC and pulse combined voltages.

Figure 22. The simulated electric field distribution of different DC amplitudes and +25 kV pulse combined voltages from the needle tip to the ground electrode

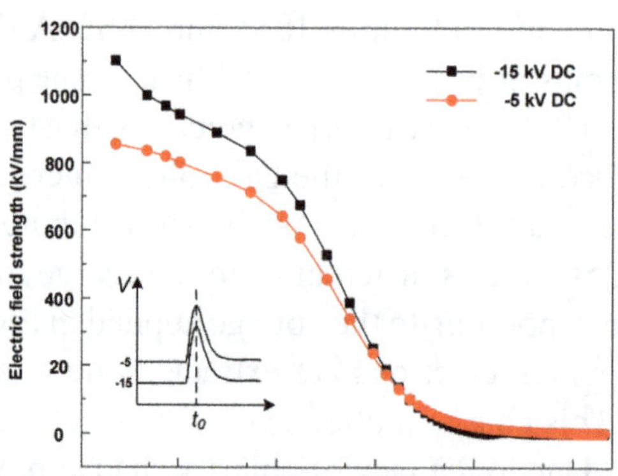

Figure 23. Relationship between the accumulated damage and the treeing time of different DC amplitudes and +25 kV pulse combined voltages

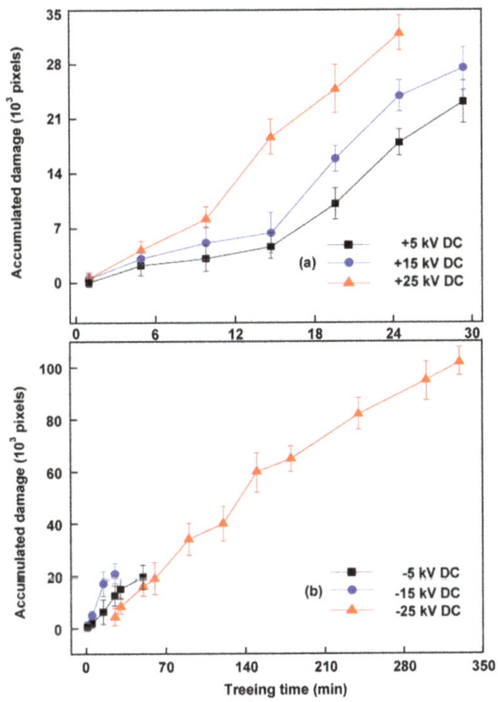

Figure 24. Breakdown time of different DC amplitudes and +25 kV pulse combined voltages

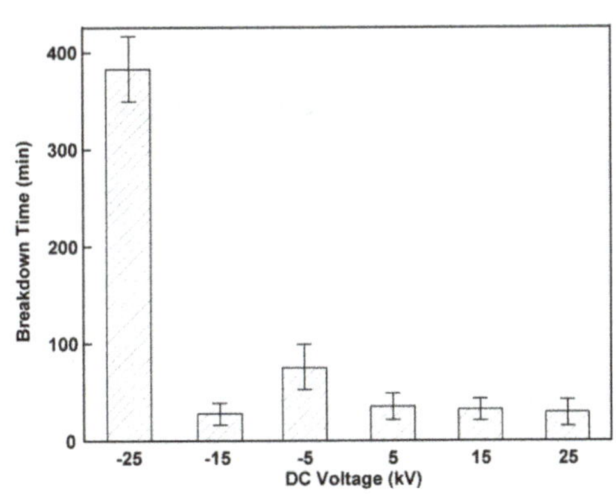

Figure 25. Typical electrical tree structures with different pulse frequencies: (a1-a3) +5 kV DC and +25 kV pulse combined voltages; (b1-b3) -15 kV DC and +25 kV pulse combined voltages

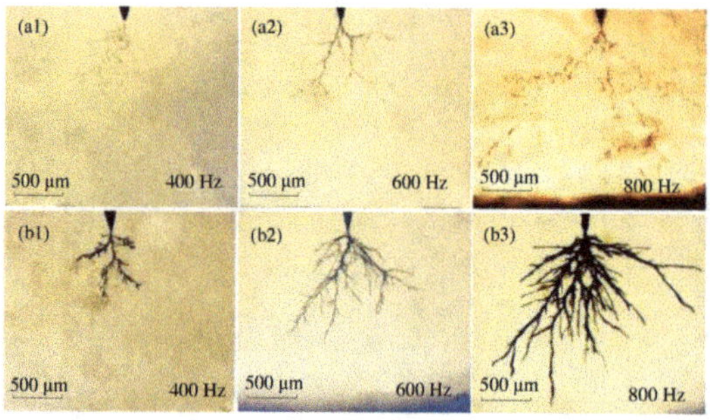

Effect of DC Voltages Combined With Pulse Voltage with the Different Frequencies

Figure 25 shows a typical tree structure of DC and pulse combined voltages with different pulse frequencies. Figures 25a1, 25a2, and 25a3 are +5kv DC and +25kv pulse combination voltages. Figures 25b1, 25b2 and 25b3 combine to a -15kv DC and +25kv surge voltage. Planting time is 10 minutes. As shown in Figure 4.8, as the frequency increases, the tree channel becomes wider and darker. The reason for this phenomenon will be explained later in this article.

Figure 26 shows the relationship between the tree length of the DC and pulse combination voltage and the tree time at different pulse frequencies. Figure 27 shows the relationship between cumulative damage at different frequencies and the tree time of the DC voltage and pulse combination voltage. The voltages in Figures 26a and 27a are +5kv dc and +25kv pulse combination voltages. The voltages of Figures 26b and 27b are a combination of -15kv DC and +25kv pulses. As shown in Figures 26 and 27, the growth rate and cumulative damage increase as the pulse frequency increases. The discharge propagates along the tube. As the pulse frequency increases, the ionization number nt in time t increases. Therefore, the damage caused by each avalanche zone is simultaneously increased, resulting in a wider branch and a greater cumulative damage.

Figure 26. Relationship between the tree length and the treeing time with different frequencies: (a) +5 kV DC and +25 kV pulse combined voltages; (b) -15 kV DC and +25 kV pulse combined voltages

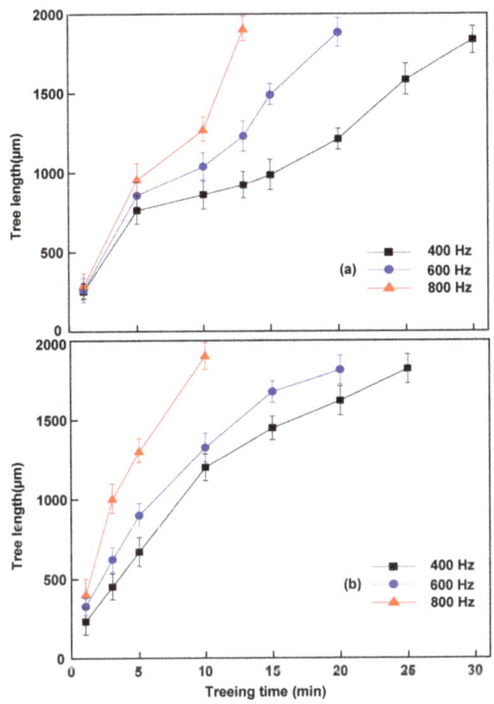

Effects of DC Voltage Combined With the Different Pulse Voltages

Figure 29 shows the waveforms of the DC voltage and the pulse voltage. Figures 29A1, 29A2, and 29A3 represent a combination of -25 kV DC and -15, -25, and -35 kV surge voltages, respectively. Figures 29b1, 29b2, and 29b3 show a combination of -25 kV DC and +15, +25, and +35 kV surge voltages, respectively. The pulse frequency is 400 Hz.

Figure 30 shows a typical tree structure with different pulse amplitudes and a combined voltage of -25 kV DC. Planting time is 15 minutes. The pulse amplitude affects the structure of the tree. As the voltage increases, the depth of charge injection will penetrate into the dielectric. As a result,

Figure 27. Relationship between the accumulated damage and the treeing time with different frequencies: (a) +5 kV DC and +25 kV pulse combined voltages; (b) -15 kV DC and +25 kV pulse combined voltages

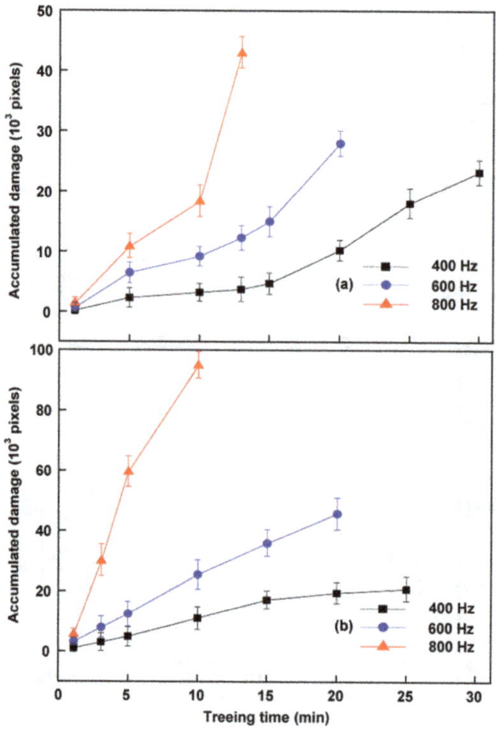

the resulting charge front is away from the tip or main branch. The wire tree behind the charging front is easier to branch. Therefore, as the amplitude increases, more and more side branches on the main branch.

Figure 31 shows the relationship between tree length and tree time for different pulse amplitudes and -25 kV DC combined voltage. Figure 32 shows the relationship between cumulative damage and tree time for different pulse amplitudes and combined voltages of -25 kV. As shown in Figures 31 and 32, the growth rate and cumulative damage increase as the pulse voltage increases. As shown in Figure 33, after the application of the negative pulse voltage, the electric field at the tip instantaneously increases during the rising phase of the pulse voltage. Once the critical field ec is exceeded, the charge

Figure 28. Breakdown time of DC and pulse combined voltages with different frequencies

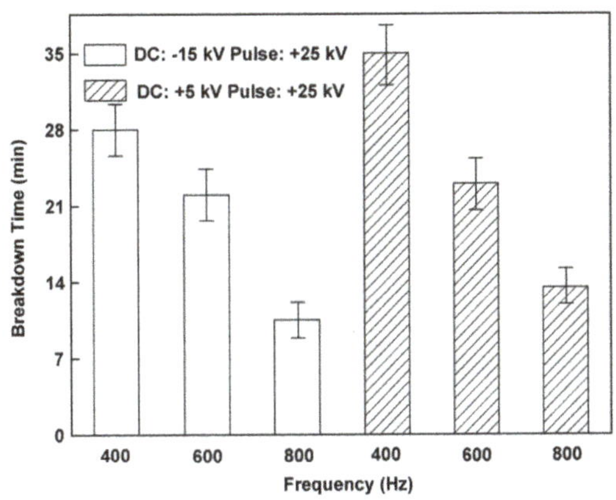

Figure 29. Waveforms of DC and pulse combined voltages

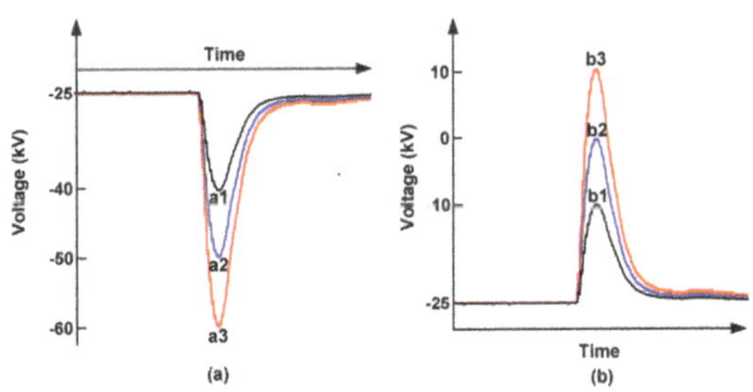

carriers will transition from a low energy state to a high mobility state and rapid charge carrier transfer will occur. Therefore, the amount of space charge injected is instantaneously increased. During the pulse voltage reduction phase, the amount of space charge injected decreases due to the decrease in tip potential. For positive pulse voltages of +15kv and +25kv, electrons will be extracted during the rising phase of the pulse voltage. During the falling phase of the pulse voltage, electrons are injected again. Since Δq+25>Δq+15, the amplitude increases from +15kv to +25kv, accelerating the tree process.

Figure 30. Typical electrical tree structures of different pulse amplitudes and -25 kV DC combined voltages

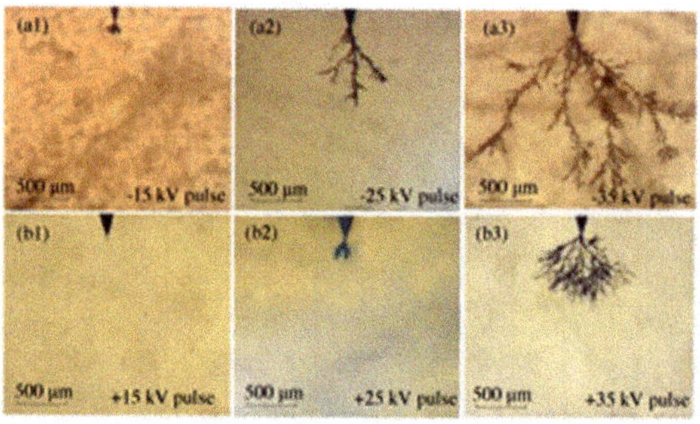

Figure 31. Relationship between the tree length and the treeing time of different pulse amplitudes and -25 kV DC combined voltages

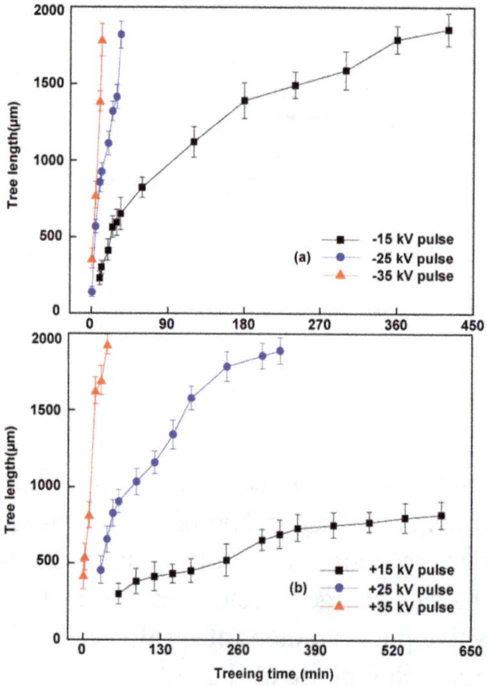

Figure 32. Relationship between the accumulated damage and the treeing time of different pulse amplitudes and -25 kV DC combined voltages

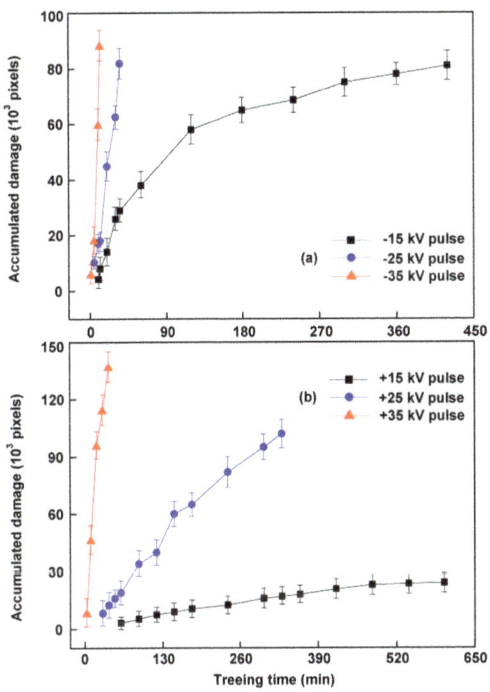

Figure 33. Schematic of the amount of injected space charges with the change of pulse voltage

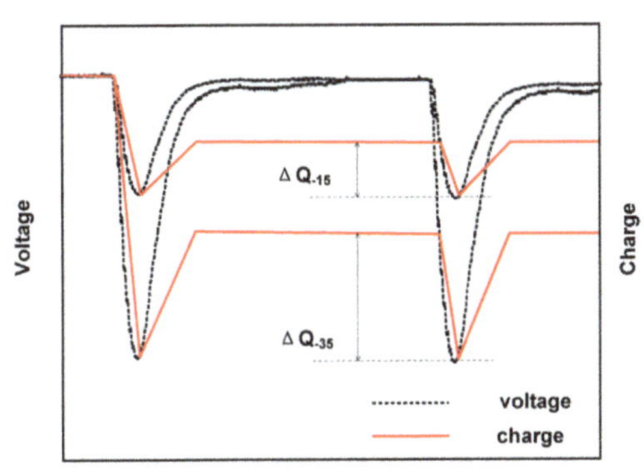

When the pulse amplitude is +35kv, some electrons are first extracted during the rising phase of the pulse voltage. When the polarity changes from the negative electrode to the positive electrode, a positive charge is injected. These positive charges combine with electrons to regenerate energy. In addition, the electric field distortion caused by space charge also accelerates the growth process of trees. As the pulse voltage increases from +25kv to +35kv, in addition to extracting charge, there is a combination of electrons and positive charges and electric field distortion, which accelerates the growth process of trees. As a result, growth rate and cumulative damage increase.

Figure 34 shows the breakdown times for different pulse amplitudes and -25 kV DC combined voltages. At a combined voltage of +15 kV pulse and -25 kV DC, there is no breakdown within 10 hours. The average effective value of the negative pulse voltage composite voltage is greater than the positive pulse voltage. As the electric field increases, the number of ionization increases. Note that for a +35kv pulse, the polarity of the voltage changes. Therefore, the electric field also increases significantly as the voltage changes from negative to positive (Zhu et al., 2019). However, the breakdown time of the -35kv pulse is less than the +35kv pulse. This may be due to the fact that the effective electric field changes more than the polarity.

Figure 34. Breakdown time of different pulse amplitudes and -25 kV DC combined voltages

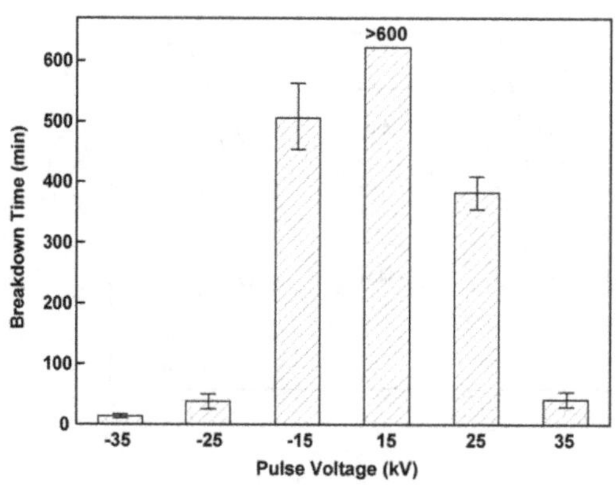

CONCLUSION

The experimental results show that the generation and growth of trees in pp are difficult. Low temperature can inhibit the growth of trees, while higher frequencies and amplitudes can accelerate the growth of trees at -30° and -196°C, but at -90°C, High amplitudes do not accelerate the growth of trees. High temperature and pulse amplitude have a great influence on the growth process of trees. As the temperature increases from 50°C to 110°C, the growth rate of trees increases. 130°C is lower than 110°C. At the same high temperature, as the pulse amplitude increases, the growth rate and fractal dimension of the tree increase. The charge transfer process is different under positive and negative DC voltages. Therefore, the effect of positive and negative DC voltage on breakdown time is different. As the positive DC voltage increases, the energy generated by the charge motion increases. Therefore, the failure time is shortened. As the negative DC voltage increases, the breakdown time decreases first and then increases. This phenomenon is caused by the uniform space charge, resulting in electric field distortion. The pulse amplitude has an important effect on the tree. As the amplitude increases, the growth rate and cumulative damage increase. A pulse voltage having the same polarity of the DC voltage is more likely to cause breakdown than a pulse voltage having a polarity opposite to the DC voltage.

REFERENCES

Bozzo, R., Gemme, C., & Guastavino, F. (1995). The effects of temperature on the tree growth phenomena and relevant PD. In *Conference on Electrical Insulation & Dielectric Phenomena, Report*. IEEE. 10.1109/CEIDP.1995.483578

Campus, A., Carstensen, P., Farkas, A. A., & Meunier, M. (2002). Chemical defects and electron trapping relevant to cable dielectrics. In *Electrical Insulation and Dielectric Phenomena, 2002 Annual Report Conference on*. IEEE.

Cavallini, A., Conti, M., Montanari, G. C., Arlotti, C., & Contin, A. (2004). Pd inference for the early detection of electrical treeing in insulation systems. *IEEE Transactions on Dielectrics and Electrical Insulation*, *11*(4), 724–735. doi:10.1109/TDEI.2004.1324362

Chen, G., Hao, M., Xu, Z., Vaughan, A., Cao, J., & Wang, H. (2015). Review of high voltage direct current cables. *CSEE Journal of Power and Energy Systems*, *1*(2), 9–21. doi:10.17775/CSEEJPES.2015.00015

Chen, G., & Tham, C. . (2009). Electrical treeing characteristics in xlpe power cable insulation in frequency range between 20 and 500 hz. *IEEE Transactions on Dielectrics and Electrical Insulation, 16*(1).

Chen, X., Xu, Y., Cao, X., Dodd, S. J., & Dissado, L. A. (2011). Effect of tree channel conductivity on electrical tree shape and breakdown in xlpe cable insulation samples. *IEEE Transactions on Dielectrics and Electrical Insulation*, *18*(3), 847–860. doi:10.1109/TDEI.2011.5931074

Danikas, M., & Tanaka, T. (2009). Nanocomposites-a review of electrical treeing and breakdown. *IEEE Electrical Insulation Magazine*, *25*(4), 19–25. doi:10.1109/MEI.2009.5191413

Du, B. X., Xing, Y. Q., Jin, J. X., Blackburn, T. R., Grantham, C., Phung, B. T., & ... . (2014). Characterization of partial discharge with polyimide film in considering high temperature superconducting cable insulation. *IEEE Transactions on Applied Superconductivity*, *24*(5), 1–5. doi:10.1109/TASC.2014.2349155

Du, B. X., Zhu, L. W., & Han, T. (2016). Effect of low temperature on electrical treeing of polypropylene with repetitive pulse voltage. *IEEE Transactions on Dielectrics and Electrical Insulation*, *23*(4), 1915–1923. doi:10.1109/TDEI.2016.7556462

Du, B. X., Zhu, L. W., & Han, T. (2016). Effect of low temperature on electrical treeing of polypropylene with repetitive pulse voltage. *IEEE Transactions on Dielectrics and Electrical Insulation*, *23*(4), 1915–1923. doi:10.1109/TDEI.2016.7556462

Gao, Y., Chen, Q. G., Chi, M., & Lin, L. (2013). Flow electrification characteristics of oil-pressboard insulation under polarity reversal electric field. In *2013 IEEE Conference on Electrical Insulation and Dielectric Phenomena (CEIDP 2013)*. IEEE.

Gulski, E., Cichecki, P., Wester, F., Smit, J., Bodega, R., Hermans, T., ... De Vries, F. (2008). On-site testing and pd diagnosis of high voltage power cables. *IEEE Transactions on Dielectrics and Electrical Insulation*, *15*(6), 1691–1700. doi:10.1109/TDEI.2008.4712673

Han, & Kjellqvist. (2012). Evaluation of organic peroxide decomposition byproducts from incompletely crosslinked high voltage power cables. In *Electrical Insulation & Dielectric Phenomena*. IEEE.

He, Z., Li, J., Jiang, T., Bao, L., & Cheng, C. (2013). Partial discharge characteristics influenced by different temperatures under pulsating DC voltages. In *Solid Dielectrics (ICSD), 2013 IEEE International Conference on*. IEEE. 10.1109/ICSD.2013.6619706

Holto, J., & Ildstad, E. (2010). Electrical tree growth in extruded s-polypropylene. In *IEEE International Conference on Solid Dielectrics*. IEEE.

Lan, L., Wu, J., Yin, Y., Li, X., & Li, Z. (2014). Effect of temperature on space charge trapping and conduction in cross-linked polyethylene. *IEEE Transactions on Dielectrics and Electrical Insulation*, *21*(4), 1784–1791. doi:10.1109/TDEI.2014.004261

Murata, Y., & Kanaoka, M. (2006). Development History of HVDC Extruded Cable with Nanocomposite Material. *Properties and applications of Dielectric Materials, 2006. 8th International Conference on*. 10.1109/ICPADM.2006.284214

Okubo, H., Kojima, H., Endo, F., Sahara, K., Yamaguchi, R., & Hayakawa, N. (2008). Partial discharge activity in electrical insulation for high temperature superconducting (hts) cables. *IEEE Transactions on Dielectrics and Electrical Insulation*, *15*(3), 647–654. doi:10.1109/TDEI.2008.4543100

Shimizu, N., Shibata, Y., Ito, K., Imai, K., & Nawata, M. (2000). Electrical tree at high temperature in XLPE and effect of oxygen. In *Electrical Insulation and Dielectric Phenomena, 2000 Annual Report Conference on*. IEEE. 10.1109/CEIDP.2000.885293

Stone, G. C., Van Heeswijk, R. G., & Bartnikas, R. (1992). Electrical aging and electroluminescence in epoxy under repetitive voltage surges. *IEEE Transactions on Electrical Insulation*, *27*(2), 233–244. doi:10.1109/14.135595

Suzuki, H., Nozomu, A., Miyake, H., Tanaka, Y., & Maeno, T. (2013). Space charge accumulation and electric breakdown in XLPE under DC high electric field. In *2013 IEEE Conference on Electrical Insulation and Dielectric Phenomena (CEIDP 2013)*. IEEE.

Zhou, Y., He, J., Hu, J., Huang, X., & Jiang, P. (2015). Evaluation of polypropylene/polyolefin elastomer blends for potential recyclable hvdc cable insulation applications. *IEEE Transactions on Dielectrics and Electrical Insulation*, *22*(2), 673–681. doi:10.1109/TDEI.2015.7076762

Zhu, L. W., Du, B. X., Su, J. G., Han, T., & Danikas, M. G. (2019). Electrical treeing initiation and breakdown phenomenon in polypropylene under dc and pulse combined voltages. *IEEE Transactions on Dielectrics and Electrical Insulation*, *26*(1), 202–210. doi:10.1109/TDEI.2018.007454

Chapter 8
Surface Charge Property of SiR/SiC Composites with Field-Dependent Conductivity

ABSTRACT

An electrical field distorted by the complicated cable accessory structure and non-uniform temperature distribution is a significant threat to high voltage direct current (HVDC) cable. Thus, the field grading material (FGM) with nonlinear conductivity can uniform local field receives attention. This chapter focuses on the surface charge property of SiR/SiC composites effected by temperature. Field strength and SiC content have a positive effect on the increase in conductivity. When the temperature increases, the threshold field decreases. At high SiC content, this phenomenon is more obvious. The influence of temperature is considered under DC voltage and impulse superimposed DC voltage.

DOI: 10.4018/978-1-5225-8885-6.ch008

INTRODUCTION

HVDC cable attachments are very important for the stable operation of HVDC transmission (Zhou, 2015). They are easier to be damaged than other components of the HVDC transmission system (Lan, 2013; Li, 2017). The reason is that the structure of the high-voltage DC cable accessory is complicated, and the insulating layer has a multi-layer structure. These characteristics lead to serious electric field distortion, which exacerbates the charge accumulation. In history, the solution to the problem has been based on dielectric constant and resistance (Virsberg, 1967; Nelson, 1984). At present, solutions based on dielectric constant and resistivity can be seen in many high voltage DC cable accessories. These solutions exist in stress cones of nonlinear conductive layer. Many studies have shown that nonlinear conductivity materials have advantages in changing local fields (Donzel, 2011; Boggs, 2015; Wang, 2012). It shows that the inorganic filler could provide high conductivity under high electric field, thereby accelerating the dissipation of interfacial charge. Based on this technology, ABB successfully fabricated a prototype 300 kV DC cable by applying FGM in the connection structure (Jacobson, 2006).

However, the cable attachment structure is not the only cause of high frequency failure. Temperature has a great effect on the electrical conductivity, which is a major factor in the field distribution under DC voltage. The HVDC cable is designed to operate at 90°C, and the conductivity of the insulating material is significantly improved (Hjerrild, 2001). The problem is that the conductivity of the insulating material increases in a different range. The conductivities of most widely used insulating materials may differ by an order of magnitude, which will certainly affect the local field (Vu, 2015). There is also a temperature gradient in the HVDC cable. The temperature in the inner insulation of the cable is high, while the temperature of the outer insulation is not high, which causes the cable to exhibit an oblique electric field from center to outside.

Another accessory failure reason to be aware of is the transient electric field. The transient overvoltage caused by the lightning and switch is inevitably generated in the HVDC transmission system. Other study has shown that lightning voltage can effect the accumulation of charge and further effect the fault beginning, such as branches. Equally important is to explain how nonlinear conductive insulation works in this situation.

This chapter aims to identify surface charge behaviors that take into account temperature and transient voltages as well as nonlinear conductive insulation. The silicon rubber (SIR) matrix is doped with SiC particles to realize nonlinear conductivity. The different amounts of SiC leads to different nonlinear conductance. Surface charge characteristics were measured by surface potential decay (SPD) under different temperatures. In this chapter, the charge characteristics of SiR/SiC materials under DC voltage and DC superimposed pulse voltage are studied.

NONLINEAR CONDUCTIVITY AND CARRIER MOBILITY OF SIR/SIC COMPOSITES

The Calculation of Carrier Mobility

Basic research begins with the preparation of SiR/SiC composites from micron-sized SiC particles and SiR matrix. Based on SPD measurements, a mcthod of obtaining carrier mobility can be established (Perlman, 1976). It is assumed that the decay loop is open in the decay process. The surface potential starts to decay when the time t is 0. The corresponding ISP absolute value is V0. *l* is the sample thickness. Besides, considering Poisson Equation and continuity equation of the current, it is derived that

$$\frac{dV(x,t)}{dt} - \frac{\rho(x,t)\mu(t)}{\varepsilon}V(x,t) + \frac{1}{2}\mu(t)E^2(L,t) = 0 \tag{2-1}$$

where ρ is the charge density; ε is the dielectric constant of the sample; μ is the carrier mobility; E is the electric field; x is the distance from the upper surface to the integration. During SPD process, the charges typically decay faster first, showing that the charges take less time to migrate. Therefore, the average moving time tT of the charge throughout the process is defined to simplify the model. When t<tT,

$$\left|\frac{dV_s(t)}{dt}\right| = \frac{1}{2}\mu(t)\left(\frac{V_0}{L}\right)^2 \tag{2-2}$$

where Vs is the surface potential. When t > tT,

$$V_s(t) = \int_0^L E(x,t)dx = \frac{L^2}{2\mu t} \tag{2-3}$$

Equation (2-2) and Equation (2-3) can be derived to average carrier mobility.

$$\mu = \frac{L^2}{t_T V_0} \tag{2-4}$$

For micron-sized-thickness samples, the SPD also obtains an approximate body trap distribution. (Llovera, 2004). SPD is considered to be caused by charge trapping and subsequent movement to ground (Das-Gupta, 1990). In particular, if chargeneutralization does not occur and injected charges can all move to the ground, the relationship between trap level and density can be determined as follows (Simmons, 1973),

$$\frac{N_1(E)}{E_1} = \frac{4_\varepsilon}{eL^2k^2T^2\ln(vt)}\left|t\frac{dV_2(t)}{dt}\right| \tag{2-5}$$

where Nt is the trap density and Et is the trap level. e, ν and k are respectively the electron volt, electronic escape frequency and Boltzmann constant.

Nonlinear Conductivity

Our previous studies have shown the conductivity of SiR/SiC composites at room temperature. As the content of SiC increases and the field increases, electrical conductivity increases. The curve of the conductivity clearly shows the threshold field. The threshold fields of M30 and M50 samples were significantly lower than M0 and M10 samples. No SiC particles were mixed in M0, and the particles were significantly dispersed in M10 due to the low content. Therefore, it is difficult to form a conductive path in a composite block (Gildenblat, 1987). Space charge limiting currents (SCLC) can cause high threshold fields in low levels of sample. (Castellon, 2011). Percolation theory shows that when the volume fraction of SiC particles is above the percolation threshold, a conductive path is formed (Hedvig, 1977). Thus, for the M30 and M50 samples, the high content of SiC is higher than percolation threshold, shortening the SiC particles distance and allowing charges to move at low fields (Ando, 1987). Furthermore, with the increasing SiC content, the conductive particles result in a high-splitting field of the SiR phase. The results show that electrical conductivity of the pure sample increases little as the electric field increases. Therefore, the large increase of conductivity in SiC-doped samples is primarily caused by injection of SiC particles into the carriers of the SiR phase.

The experiments show that the conductivities at higher temperatures have similar trend with room temperature. Table 1 shows the variation of the threshold field at different temperatures of the SiR/SiC composites. For the M30 and M50 samples, the threshold field drops more and becomes less than 1 kV/mm. The threshold field becomes lower for samples with low SiC content at 70°C and 90°C. The rate seems to increase immediately. The high temperature causes the threshold field of the composite to drop greatly, and this phenomenon is more pronounced when the SiC content is high. The intrinsic carrier concentration of SiC accords with the following equation (Hu, 2010):

$$n_i = (N_C N_v)^{1/2} \exp(-\frac{E_g}{2k_0 T}) \tag{2-6}$$

where ni is the intrinsic carrier concentration; Nc and Nv are respectively the effective state density in the conduction band and valence band; eg is the forbidden band width; k0 is the Boltzmann constant at absolute zero. The amount of effective charge in the composite increases with increasing intrinsic carriers within the conduction band. High temperatures provide carriers with greater energy to trap and jump obstacles, which increase the macroscopic decline of the threshold field even if the electric field is low.

To investigate the nonlinear conductance characteristics at high electric fields, temperatures range from 30°C to 90°C. According to previous research data, the conductivity increases with electric SiC, temperature and field. Figure 1a shows the conductivity of the M0 sample at different temperatures.

Table 1. Threshold field of the conductivities of SiR/SiC composites at different temperatures

	M0	M10	M30	M50
RT	20 kV/mm	17 kV/mm	3.8 kV/mm	1.1 kV/mm
30°C	14 kV/mm	10.1 kV/mm	1.7 kV/mm	<1 kV/mm
50°C	9.6 kV/mm	5.5 kV/mm	<1 kV/mm	<1 kV/mm

Figure 1. The conductivities of the composites under different electric field and temperatures

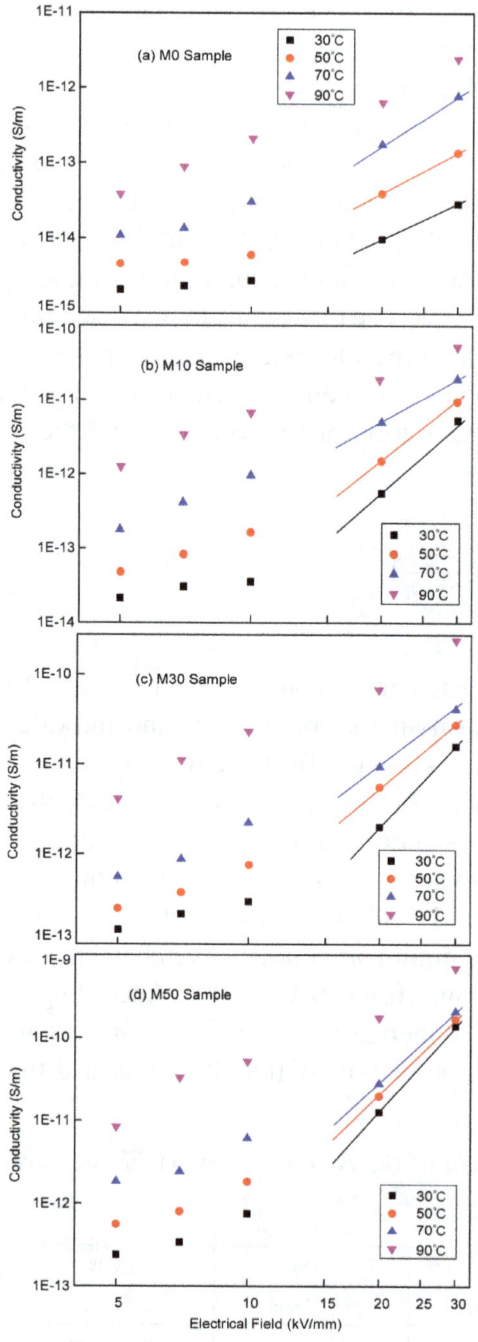

Regardless of the strength of the magnetic field, the rate of increase in conductivity increases at high temperatures. The conductivity at different temperatures continues to increase and divergence at high temperatures. However, when the temperature was raised to 90°C, no convergence was observed. Figures 1c and 3d show the same phenomenon as Figure 1b. It can be reasonably assumed that this phenomenon is caused by SiC particles due to the consistent of other experimental factors. A simplified model shown in Figure 2 can well explain the phenomenon. The conductive path of the entire composite block can be analyzed as follows. The first part is the interior of the silicon carbide particles, as shown in Figure 2a. It is assumed that the silicon carbide particles have a perfect crystal structure and no impurities. Electrons are easily excited because of the narrow band gap of the sic covalent bond. These electrons cross the forbidden band and enter the conductive gap, becoming an intrinsic carrier. However, a high purity semiconductor is required in order to observe the intrinsic excitation only. In fact, ordinary semiconductors with certain impurities and crystal defects also provide carriers by ionization. At relatively low temperatures, a large number of carriers are generated by impurity ionization, while the intrinsic excitation is not significant. When the ionization is completed, the carrier concentration reaches the table value. As shown in Figure 2b, the second part of the conducting routes is from SiC partlets to SiR matrix ((Tanaka, 2005). There is an interface barrier between them. At the same time, the temperature and electric field accelerate the increase in conductivity. While the conduction process is limited by two factors for other samples. One assumption is to follow. When the temperature is lower than T2, the dry weight concentration and the chronic increase of the transition area. As shown in Figure 2b, the fence is tilted for the relatively low field E1. And only a part of the carrier can jump on the fence and move through the compound cannon. The dam inclination is caused by E2, which is sufficient to cause the largest carrier to jump on the dam, which means that the effect of the high field is saturated at a relatively low temperature. Since the high conductivity, the conduction convergence of the SiC-doped samples occurs only in the case of a high field if the temperature is 50-70°C. Similarly, in the case where 90°C may exceed T2, the carrier concentration in the inner region is one. There is an insufficient rapid increase to make most carriers over the barrier even the electric field is 20-30 kV/mm, with the amount of hopping carriers continuing to increase. Therefore, the conductivity is not limited and convergence is not present at 90°C. However, this theory could be applied to nanocomposites only, with a

Figure 2. Conduction schematic of SiC particles and SiR/SiC composites

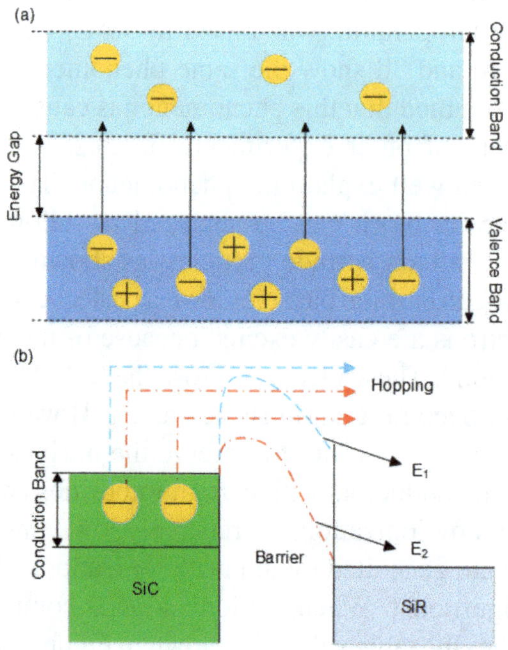

Figure 3. Schematic of the relation between the carrier concentration and the temperature in the impure semiconductor

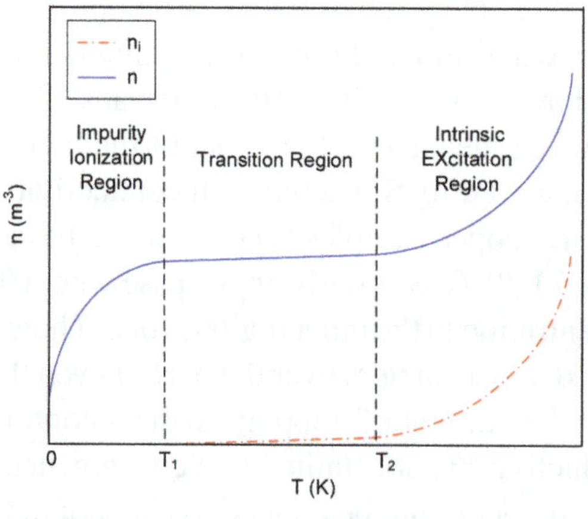

wide interface area. While the interface groove may not be able to transport during carrier transport for a micro-filler with a low surface-to-volume ratio.

Trap Characteristics

The trap characteristics of the composites materials have a close connection with the conductivity. As a typical example, Figure 4 shows the trap distribution at different temperatures at -6 kV. The bimodal distribution of all samples was evident at all temperatures. When the temperature was 30°C, the M0 sample had a shallow trap peak of 0.86 eV and a deep trap peak of 0.95 eV. High temperatures cause the traps trapped by deep traps to jump out of the SPD process and participate in charge transport, detecting more charge in deep traps. Therefore, as the temperature rises, both peaks move toward a high energy level. As shown in Figure 5, when the temperature was raised from 30° C to 90° C, the deep trap peak of the sample m0 decreased from 4.05 × 1020 to 3.11 × 10 20 eV -1·m -3.

At the same time, the peak value of the shallow trap increased from 1.90×1020 to 2.75×1020 eV-1·m-3. At high temperatures, high-energy carriers tend to stay in shallow wells, easily escaping, resulting in a large amount of moving charge on the microscopic and macroscopically higher conductivity. These trends are usually still present in the sic mixed sample. The sample M10 predominates at 90°C and sample M30 at 70°C. The peak value of the shallow trap in sample M50 at 50°C is almost the same as the peak value of the deep trap, as shown in Figure 2.5. As shown in Figure 1, these two causes lead to higher conductivity of the SiC mixed sample.

Carrier Mobility

Using the data obtained by the SPD experiment, plot the relationship between lg dvs(t)/dt and lgt, and it is possible to find tT. Besides, at higher temperatures, the increase in mobility is more pronounced, which can be explained by equation (2-6) and Figure 2. Based on the results of SPD experiment shown in Figure 6, the field strength at which the sample begins to withstand was between -11.2 and -13.8 kV/mm. Considering Figure 1, when the magnetic field is lower than 15 kV/mm, the conductivity increases more markedly at high temperatures. Therefore, the relationship between the carrier mobility and the conductivity shown in Figure 6 satisfies a positive correlation. It can also be seen from Figure 5 that the carrier mobility increases with the

Figure 4. Trap distribution of the composites charged by -6 kV voltage for 10 min

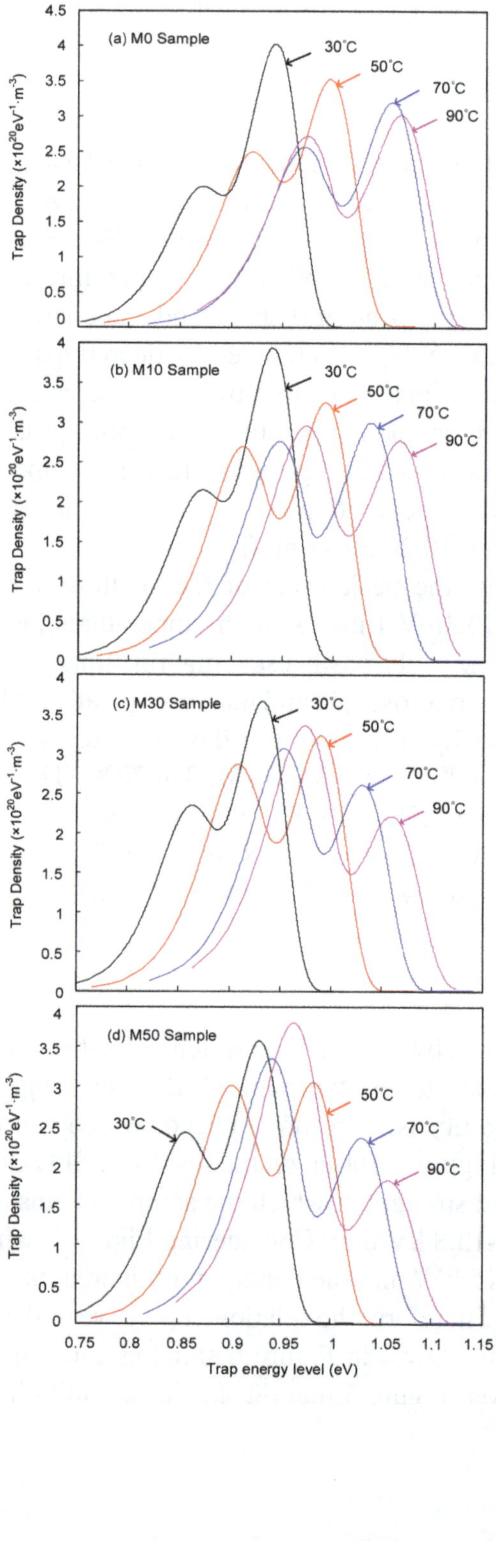

Figure 5. Shallow-trap and deep-trap peak values of M0 and M50 charged by -6 kV voltage for 10 min

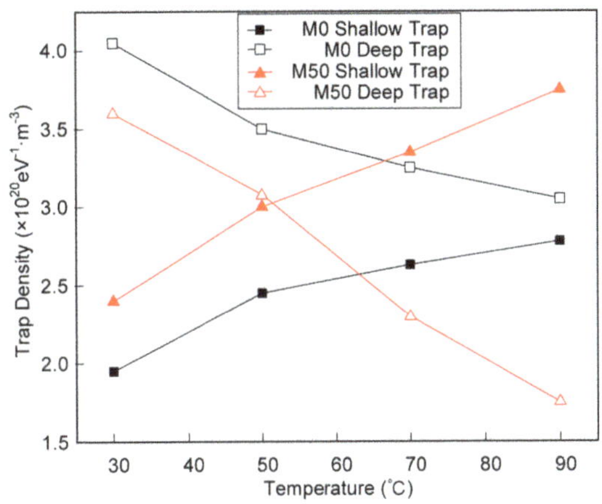

Figure 6. Carrier mobility of the composites charged by -6 kV voltage for 10 min

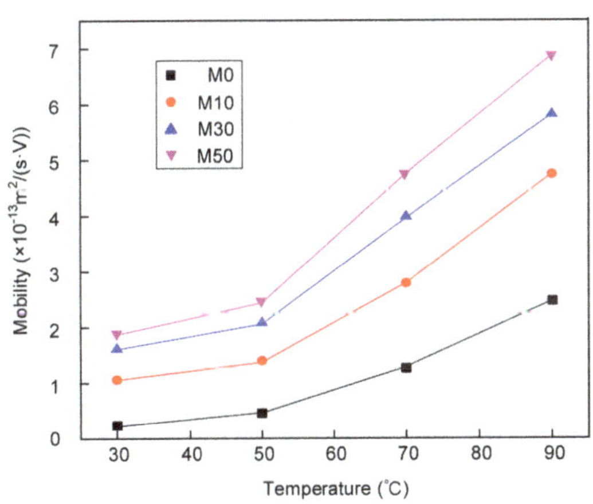

increase in shallow trap density. Therefore, shallower traps correspond to high carrier mobility and high conductivity (Du, 2018).

The carrier mobility is also effected by the applied voltage. The carrier mobility of M50 in Figure 7 is to show the difference in applied voltage and temperature. From the point of view of carrier motion, the temperature causes carriers to pass through the bulk of the composite with high thermal

Figure 7. Carrier mobility of the M50 sample under different voltages

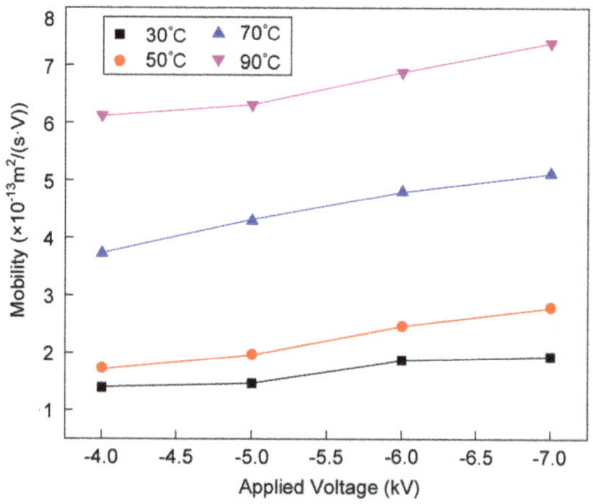

kinetic energy, while the voltage tilts the barrier and provides high potential energy. According to Figure 7, when the voltage is increased from -4 to -7 kv, the mobility increases by less than 2 x 10-13 m2·(s·V) -1 regardless of temperature. When the temperature is raised from 30° C to 90°C, the carrier mobility at different voltages is increased by 4 to 5 times. Therefore, for carrier mobility, temperature is more efficient than voltage. Interestingly, the conductivity distribution in Figure 1d is the same as the electric field below 15kV/mm. The temperature increases the conductivity by nearly two orders of magnitude, while the electric field only increases the conductivity by several times.

SURFACE CHARGE PROPERTY OF SIR/SIC COMPOSITES UNDER IMPULSE SUPERIMPOSED DC VOLTAGE

This section shows the surface charge characteristics at 30°C under repeated pulsed superimposed DC voltages. The DC voltage is set at -3, -4, -5, -6 kV, and the repetition pulse peak is set at -2, -3 kV. In order to get the transient characteristics of the surface charge, a DC voltage is applied for 5 min, and then the repetitive pulse is turned on in the last 5 s, which is referred to as DC superimposed pulse voltage. For example, -4 kV DC voltage is called D4, and a voltage formed by a -4 kV DC voltage and a -3 kV repetitive pulse is called D4P3. In order to enhance the surface charge motion, SiR/SiC composites

Figure 8. Relationship between the conductivity and the field of the composites

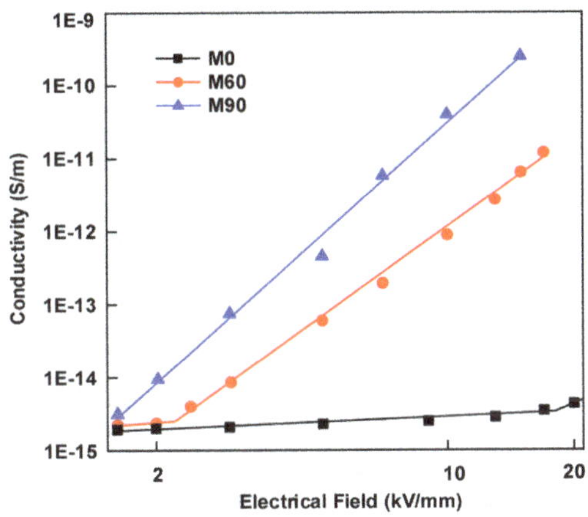

with larger particle sizes were used for the study. The figure below shows the conductivity. The M90 sample had the highest sic particle content and there was no SiC particle in the M0 sample.

Surface Charge Property of the Composites Under DC Voltage

The ISP of the composites with different SiC contents under DC voltage is shown in Figure 9. Record the 20 min decay process. The average attenuation rate of different samples within 15 s before the SPD process is shown in Table 2. Its calculation formula is as follows:

$$R = \frac{V_0 - V_{15}}{t} \tag{3-1}$$

The high ISP under -6 kV results in higher conductivity. Because of the high conductivity, the ISP decreases by more than 500 V and the residual charge decreases by nearly 800V after 3 min compared with M60. As the applied voltage increases, the ISP of M0 increases and the attenuation rate increases. However, the surface charge decay rate is not large, and the residual charge is still considerable after 20 minutes, which corresponds to the low conductivity in M0. When the content of SiC particles is increased to 60

Figure 9. Relationship between the absolute value of the ISP of the composites and the DC voltage

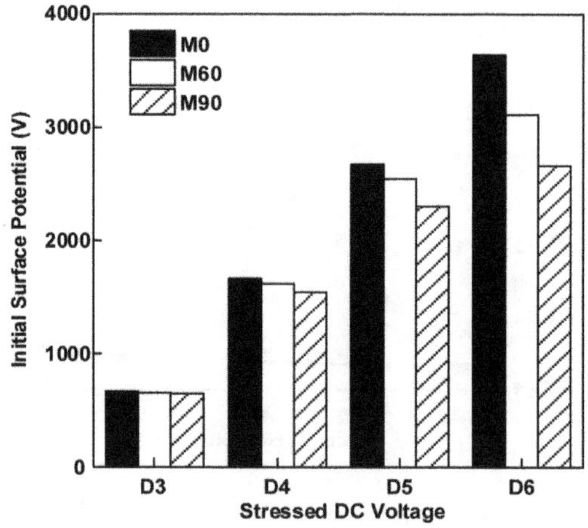

Table 2. SPD rate of the composites under DC voltage

Voltage (kV)	SPD Rate (V/s)		
	M0	M60	M90
D3	0.12	0.53	1.00
D4	0.47	3.06	13.00
D5	0.8	20.72	42.07
D6	1.52	54.68	92.27

parts, the surface charge decreases rapidly under high voltage direct current corona charging, which in turn accelerating the surface charge accumulation during corona discharging. Therefore, as the DC voltage increases, the SPD rate becomes faster. Extremely serious, the -5 kV and -6 kV surge protector data even showed crossover, as shown in Figure 10. The surface charge of the 6 kV decays to the same surface potential value as the -5kV SPD data in less than 40s due to the higher conductivity. Compared with M60, ISP decreased by more than 500 V at - 6 kV, and after 3 minutes of attenuation, the rest of the surface decreased by nearly 800 V due to high conductivity. For M90, cross phenomenon is more obvious.

Figure 10. Typical cross-over phenomenon

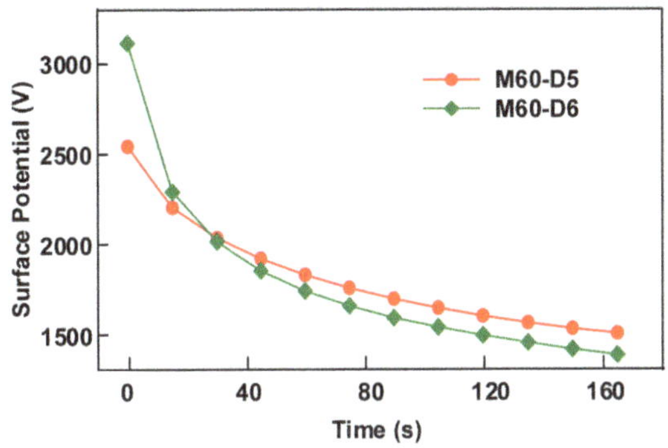

Surface Charge Accumulation of the Composites Under Impulse Superimposed DC Voltage

In order to verify the effect of repetitive pulse on the surface charge accumulation, the ISPs of Sir / SiC composite with different SiC content under the action of pulse superimposed DC voltage is shown in Figure 11. The ISP of d3p2 and d4p2 showed that the ISP of M0, M60 and M90 increased

Figure 11. Relationship between the absolute value of the ISP of the composites and the impulse superimposed DC voltage

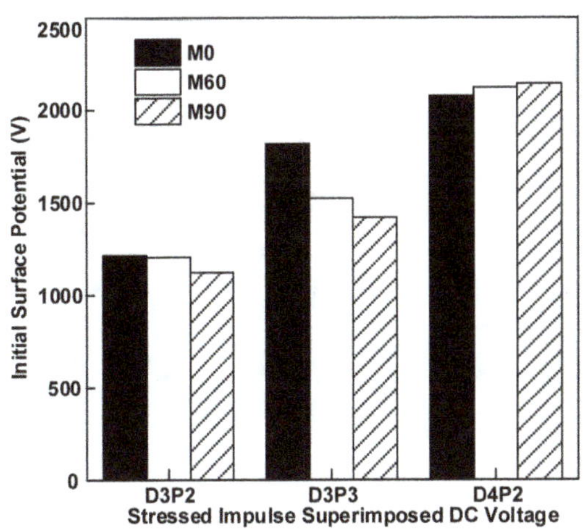

by - 869, - 921 and – 1021V, while the ISP of d3p2 and d3p3 showed that the ISP of M0, M60 and M90 increased by - 618, - 319 and - 298V. Figure 12 shows a simplified model for explaining these different trends. There are two parts that influence the ISP: horizontal diffusion and repulsion. Firstly, when the voltage increases from -3KV to -4KV, the surface potential increases at the last stage during the corona charging. With the pulse in the last five seconds, additional charge is injected into the surface. The newly arrived charge knocks the existing DC voltage charge back to the adjacent area (DU, 2017). The amount of existing DC voltage charge determines the rejection effect. Figure 9 shows that the ISP of the m90 sample at -4kV DC is the smallest, indicating its high nonlinear conductance. Therefore, the ISP of M90 samples with the weakest rejection effect increased the most. Therefore, the rejection effect of the existing DC voltage charging is not very large. For M60, the -3 kV electric field calculated is not sufficient to produce nonlinear conductance. As a result, the nonlinear conductance of M60 and M90 starts to take effect when the increase of repetitive pulse brings additional charge,

Figure 12. Formation of the ISP under impulse superimposed DC voltage

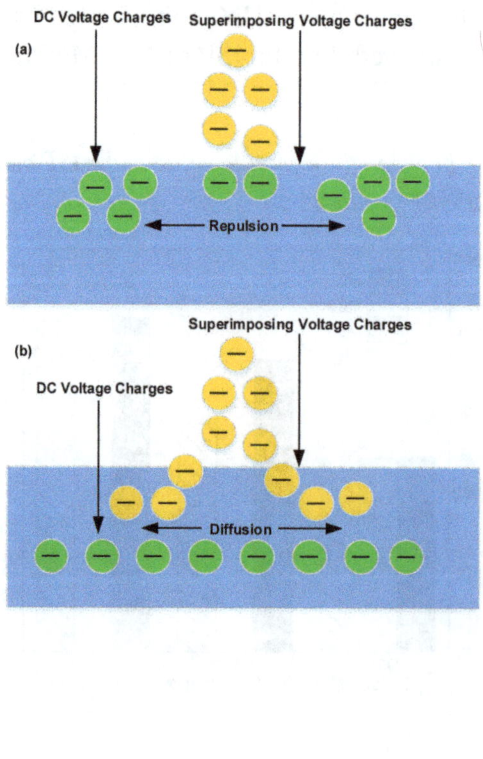

and many newly arrived charges spread to adjacent areas, as shown in Figure 12b. Therefore, the increase of ISP in M90 samples is the smallest.

Surface Charge Decay of the Composites Under Impulse Superimposed DC Voltage

For M0 samples, D3, D3P2 and D3P3 ISP shows that repeated pulses will generate more accumulated charges. As a result, the attenuation rate of these SPDs increased from 0.13 V/s to 1.79 V/s. However, the attenuation rate of M60 D3P2 is 62.7% higher than that of D3P3, that of M90 is 120.9%, and that of M0 is only 27.4%. The high surface potential caused by repetitive pulse causes the nonlinear conductivity to dissipate the surface charge immediately, and this effect is more obvious in the high SiC-doped samples. This shows that the effect of repeated pulses on nonlinear conductance is very obvious.

Interesting phenomenon occurs when comparing D3P2 and D4P2. For M0 samples, the attenuation rate increases by 35.6% with the increase of DC voltage, which is greater than the attenuation caused by the increase of repetitive pulse. The same trend was also found in the data for the SiC-doped samples. If the increase of DC voltage amplitude is the same as that of repetitive pulse voltage, the increase of DC voltage will accumulate surface charge more effectively. Therefore, the increase of DC voltage leads to higher nonlinear conductivity, which leads to higher attenuation rate. The repeated pulses do not provide enough energy for the charge. As long as there is a large amount of charge in the corona channel, as long as the pulse peak value is passed, the pulse driven injection process will end. As a result, the increase of attenuation rate caused by the increase of repetition pulse is small. This effect can be found in all samples whether or not there is nonlinear conductivity.

As mentioned above, high SPD rate results from the high ISP. However, there may be some differences between the situations. Similar phenomena occur in SiC-doped samples. The point directly below the needle electrode

Table 3. SPD rate of the composites under the impulse superimposed DC voltage

Voltage (kV)	SPD Rate (V/s)		
	M0	M60	M90
D3P2	1.46	4.26	5.53
D4P2	1.98	14.41	27.27
D3P3	1.86	6.93	12.27

on the sample is regarded as the origin. According to the distance from the center point, measuring points are named respectively, such as 1 cm points and 1.5 cm points. Figure 13a shows the surface potential of the M60 sample under D3P3. The surface potential at the origin and 0.5cm points decreases with time, and the surface potential values at these two points are very similar. In addition, there are differences in the way surface potentials change over time. Figure 14 shows the average decay rates of M60 samples of D3P3, M60 samples of D4 and M0 samples of D3P3 at 1 cm point in different time ranges. The negative decay rate indicates that the surface potential increases before 30 s, but decreases only after 60 s. The surface potential at 1 cm is not less than the ISP before 120 s. Obviously, the potential of the center surface becomes low lower whereas the surface potential of the outer region becomes higher. This is the result that the repetitive pulse cannot be eliminated immediately because it injects a large amount of charge in a short time. At the end of the charging process, these extra accumulated charges will show a significant horizontal potential difference from the center to the adjacent area. After 60 s, since most of the surface charge has decayed, the surface potential at all positions will decrease. As a comparison, figure 13b shows the surface potential of the M60 sample under D4. Before 60 s, the surface potential difference at the origin, 0.5 and 1 cm is small. Since there is no repetitive pulse, the over accumulated charge in the center disappears. In addition, the relative decay rate in Figure 14 is time independent and positive. The average attenuation rate for 120 and 300 s is higher than M60-D3P3. Because the surface potential of M60-D3P3 at 1cm can be supplemented from the central charge, thus reducing the attenuation rate. All of these clues indicate that there is no significant horizontal motion of the surface charge.

In addition, nonlinear conductivity has a large effect on charge level motion. Figure 13c shows the surface charge potential of M0 under D3P3. Similarly, compared with the central region, the surface potential at the point of 1cm decreases greatly, which indicates the overcharge caused by repetitive pulse. M0 the decay rate in Figure 14 is initially positive and becomes negative after 30 s. The horizontal movement of the surface charge is much slower than that of M60 on D3P3. The center charge is dissipated to the surroundings, so that the surface potential at 1 cm of m60 starts to increase immediately. This is because the high nonlinear conductance caused by the excessive horizontal potential difference quickly responds to the excessive accumulation of the center charge. Thus the surface potential at 1 cm point shows similar degradation

Figure 13. The absolute value of the surface potential of the composites

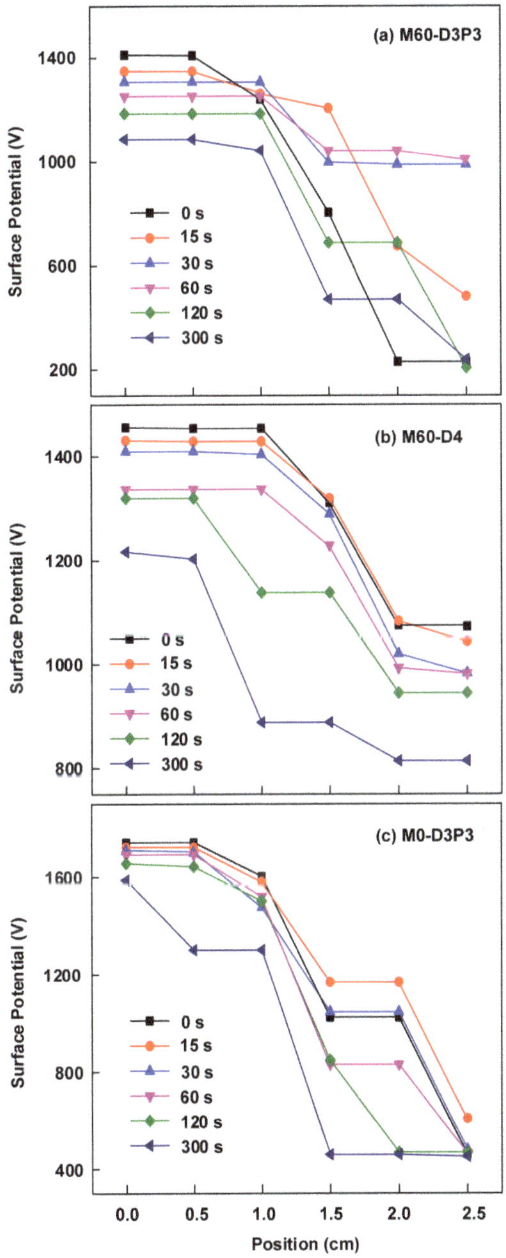

Figure 14. Average SPD rate at 1 cm in different time ranges

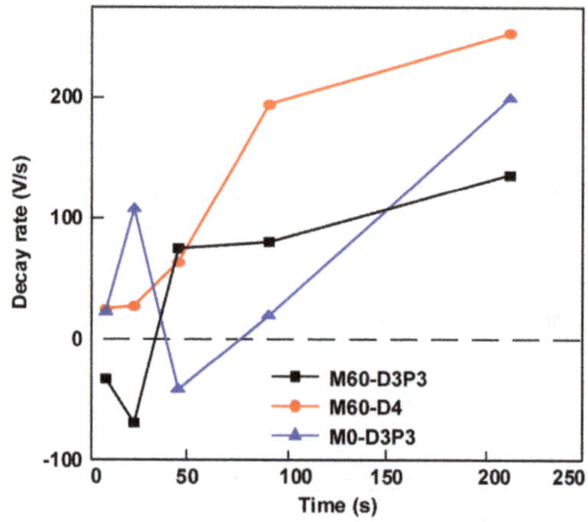

performance to that of the DC charged sample. The surface potential at the beginning of M0 is still very high, 120 s. Changes in trap and carrier mobility can also reflect the horizontal movement of charge. Table 4 lists the relevant trap characteristics and carrier mobility for the m60 samples. From the data in Table 2, Table 3 and Table 4, it can be concluded that the more shallow traps, the higher the carrier mobility and the higher the attenuation rate. In particular, the trap density and carrier mobility of D3P2 are greater than D4. This corresponds to the discussion of figures 9 and 11. From this point of view, the horizontal motion of the surface charge will lead to the equivalent shallow trap (Du, 2019).

Table 4. Trap properties and carrier mobility of the M60 sample

Voltage (kV)	Carrier Mobility ($m^2 \cdot (s \cdot V)^{-1}$)	Shallow Trap Level (eV)	Shallow Trap Density ($m^{-3} \cdot (eV)^{-1}$)
D3	1.54×10-13	0.86	1.062×1018
D4	5.27×10-13	0.86	3.842×1018
D5	7.23×10-13	0.84	1.612×1019
D6	8.83×10-13	0.82	2.472×1019
D3P2	5.56×10-13	0.84	5.562×1018
D4P2	6.89×10-13	0.84	1.162×1019
D3P3	6.23×10-13	0.84	5.842×1018

SURFACE CHARGE PROPERTY OF SIR/SIC COMPOSITES UNDER LIGHTNING IMPULSE SUPERIMPOSED DC VOLTAGE AT DIFFERENT TEMPERATURES

In this section, the standard lightning wave with high power is applied at different temperatures. The DC voltage is set to - 4 kV and the peak value of the pulse voltage is set to - 5, ±10 kV. In order to obtain the characteristics of the surface charge under DC superimposed pulse voltage, the sample will be applied by the DC voltage for 5 minutes, and the lightning voltage will be turned on in the last. For example, - 5 kV DC voltage is called D5, and the pulse superimposed DC voltage formed by - 5 kV DC voltage and - 6 kV or 8 kV pulse voltage is called D5P5 or D5-P8.

Nonlinear Conductivity Characteristics

The conductivity of the composites under different temperatures is shown in Figure 15. Figure 15a shows the conductivity at 30°C. The conductivity of M0 sample follows Ohm's law and increases slowly with the electric field. Therefore, the current is directly proportional to the electric field. When the electric field is high, the conductivity increases at a higher rate, which is caused by the SCLC. In M60 and M90 samples, the conductivity shows a higher rising speed at a lower electric field because it is caused by conductivity rather than nonlinearity of SCLC. The sic particles serve as semiconductors to provide more carriers within the conduction band. When the electric field is close to the threshold, the number of conduction band carriers increases sharply. Under high electric field, the potential barrier between SiC particles and matrix will tilt. The electric field also increases the carrier energy. In addition, according to the percolation theory, when the packing volume fraction is large enough, the packing particles can form a conductive network.

At 50, 70 and 90°C, the mixed samples of M0 and SiC are very different. For M0, the conductivity is directly proportional to the temperature compared to the conductivity at 30°C. The carriers in m0 are mostly located in the forbidden band and are hardly driven by the electric field. Only a few carriers participate in the conduction. But even if the electric field is not high, high temperature can provide more energy for these carriers to escape from traps and jump barriers. Therefore, the conductivity increases in the whole range of electric field at high temperature. Due to the high conductivity at low

Figure 15. Relationship between the conductivity of the composites and the electric field

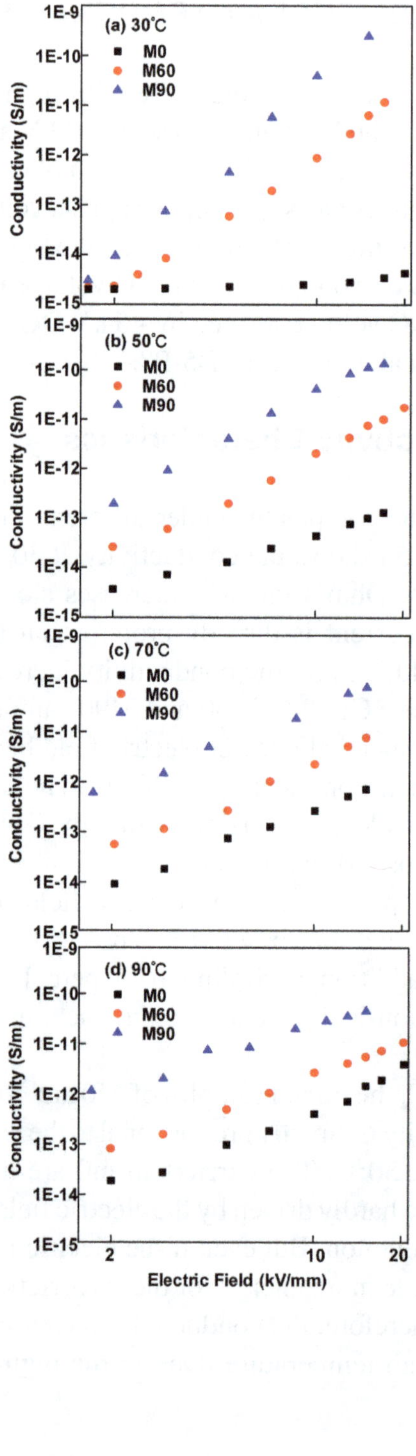

electric fields, the threshold field becomes unclear. For M60, the conductivity of more semiconducting SiC particles is higher than that of M0, because the conductivity seems to increase rapidly in a very low field. When the electric field is less than 10 kV/mm, the conductivity is similar to M0, indicating that conductivity increases with the increasing temperature. But the increase becomes negligible at 90°C. M90 occurs the same phenomenon. Figure 16 clearly shows the slope of conductivity relative to field. This indicates that temperature has a positive effect on increasing the conductivity of M0, but has a negative effect on M60 and M90. Besides, the effect of temperature is more pronounced in samples with higher SiC content. Therefore, the temperature can even play a negative role in increasing the conductivity of the sample. A typical case is shown in Figure 18. The slope of the conductivity-temperature curve is defined herein as the conductivity-temperature coefficient. At 15 kV/mm, the coefficient of M0 is positive, with M60 almost 0 and M90 negative. This phenomenon is of great significance because it may be the key to solve the insulation fault, which is caused by the uneven distribution of conductivity caused by the temperature gradient. The molecular structure of Sir is an amorphous chain. SiR acts as a good insulator and most carriers are confined within the valence band of the SiR molecule. High temperatures could exacerbate the thermal motion of the carrier, thus increasing the possibility of the stimulated carrier jumping into the conduction band. The current is formed by these excited carriers being driven by an electric field. This is why m0 has a positive conductance temperature coefficient. SiC particles are different from Sir because they are composed of crystal structure, as shown in Figure 18. The gap between the valence band and the conduction band of the semiconductor is relatively narrow. In this case, the temperature-induced excitation is not obvious. At the same time, the increase of temperature will also lead to lattice vibration. According to classical semiconductor theory, vibration is described by lattice waves including long longitudinal waves (LLAW) and long longitudinal waves (LLOW). LLAW and LLOW are used to indicate that adjacent atoms share the same vibration phase and reverse vibration phase. Lattice vibration scatters electrons in the current, resulting in a decrease in conductivity. The scattering probability is described by the following equation:

$$P_{LLAW} = c_1 T^{3/2} \tag{4-1}$$

Figure 16. Slope of the conductivity to electric field of the composites

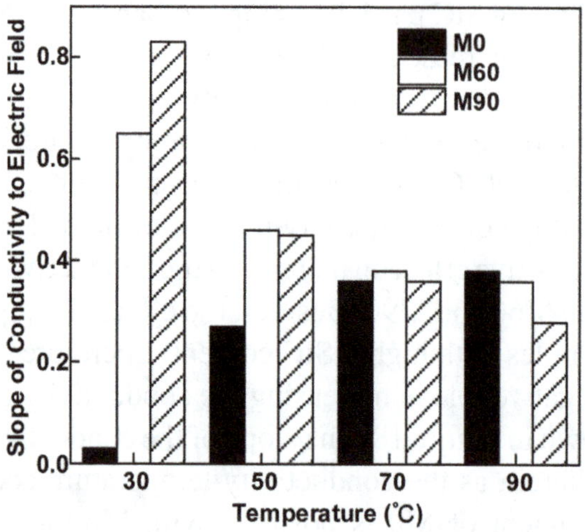

Figure 17. Relation between the conductivity of the composites and the temperature at 15 kV/mm

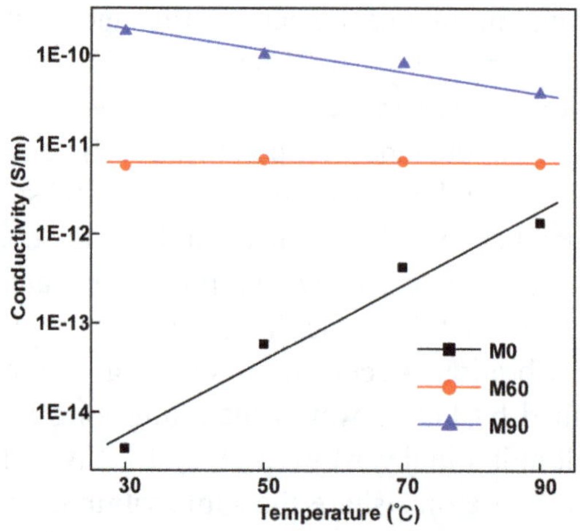

Figure 18. Structure of the SiC particles and crystals

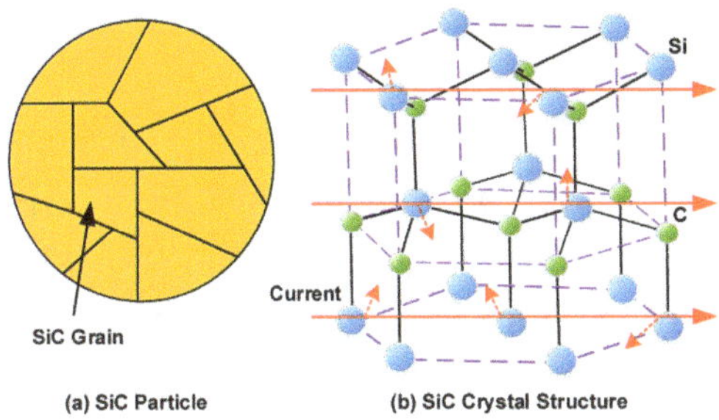

$$P_{LLOW} = c_2 \frac{(h\nu)^{3/2}}{(k_0 T)^{1/2}} \cdot \frac{1}{e^{\frac{h\nu}{k_0 T}} - 1} \cdot \frac{1}{f\left(\frac{h\nu}{k_0 T}\right)} \tag{4-2}$$

Where P_{LLAW} and P_{LLOW} are scattering probability, T is temperature, h is Planck constant, k_0 is Boltzmann constant, f is $h\nu/k_0T$ function, its value is between 0.6 and 1, and c_1 and c_2 are constant coefficients. Formula (8) clearly shows that pllaw increases with the increase of temperature. Therefore, pllow also increases with the increase of temperature I (Yang, 2016). This indicates that high temperatures increase the probability of electron scattering and more collisions between electrons and nuclei. Metal conduction theory can also be used to explain this phenomenon. Nonlinear fillers of insulating materials are often re-sintered to achieve significant nonlinearities compared to high purity semiconductors (Srumpler, 1999; Grundmann, 2010). The filler particles are composed of many filler particles, as shown in figure 18a. The results show that the purity of the SiC filler is about 99%. This results in many carriers for both SiC and impurities. Therefore, SiC particles can be regarded as degenerate semiconductors with resistance temperature characteristics similar to those of metals. Carriers are also scattered by nuclear vibrations in the metal. The following empirical formula can be used to describe their resistance:

$$\rho(T) \approx \frac{AT}{4M\Theta_d^{\,2}} (T \square \Theta_D) \tag{4-3}$$

$$\rho(T) \approx \frac{AT}{4M\Theta_d^{\,2}}(T\square\Theta_D) \tag{4-4}$$

where ρ is resistance, T is temperature, A is a metal-depended coefficient, M is the metal atomic mass, and Θ_D is the metal Debye temperature. Clearly, the resistance increases with the increasing temperature. The conductivity of m60 and m90 is further reduced because these effects significantly hinder the movement of electrons.

In addition, samples doped with 0.45 μm sic particles were tested and different phenomena were obtained (Xie, 2013). Even if the slope of conductivity decreases with respect to temperature, the conductivity becomes higher at high temperature. As shown in Figure 18b, when the carrier in the current passes through the SiC crystal structure, it is scattered due to the lattice vibration. If the content of SiC is the same, the current will be more restrained when flowing in large SiC particle. This indicates that the increase in conductivity caused by temperature is more pronounced by the inhibition of the continuous sic crystal structure.

Surface Potential Property of the Composites Under Homopolar Impulse Superimposed DC Voltage

Figure 19 shows the ISP of the composites under different temperatures under DC voltage superimposed by homopolar pulses. With the increase of pulse voltage, all samples experience the rise of ISP. With the increase of SiC content in the sample, the increase amplitude decreases, which helps to improve the conductivity, which helps to transfer the surface charge to the ground. When the pulse voltage is - 5 kV, the temperature has little effect on ISP reduction. However, when the pulse voltage increases to - 10kV, the decrease is obvious. The electric fields in the roughly calculated mixed SiC samples are all around 10 kV/mm. According to Figure 15, the conductivity is still proportional to temperature over this range of electric fields. Therefore, higher nonlinear conductivity is caused by an increase in surface potential, resulting in a significant decrease in the ISP of M0 and M60 at 90°C. This reduction rate is still not obvious for M90 at 90°C. This may be because m90 has a relatively high conductivity at temperatures below 90°C, thus fully inhibiting the accumulation of surface charge. The first 15 s decay rate was calculated to prepare for the next study.

Figure 19. ISP of the composites stressed by a homopolar impulse superimposed DC voltage

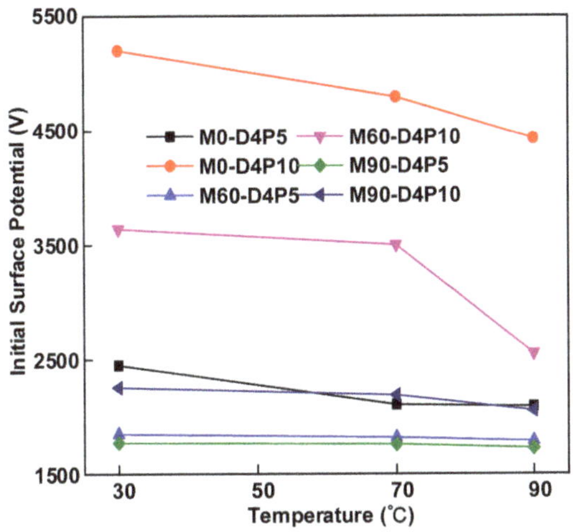

Figure 20 shows the corresponding SPD rate over the first 15 s under DC voltage superimposed by homopolar pulses. Obviously, the increase of SiC content and pulse voltage can accelerate the attenuation process. Temperature helps increase the rate of decay. An interesting phenomenon occurs that the attenuation rate of D4P10 at 90°C is even larger than M60. Therefore, the corresponding SPD rate is more sensitive to conductivity. It should also be noted that SPD is actually also affected by many factors other than conductivity. This interpretation serves as a hypothesis as to whether this phenomenon is uncertain in experiments with similar materials.

Table 5 shows another interesting comparison. The results showed an increase in ISP at different temperatures between D4P5 and D4P10. In general, when the voltage rise is the same, the surface charge suppression effect is

Table 5. Increase in the ISP between the D4P5 and D4P10 stressed sample

Temperature /(°C)	Increase in ISP /(V)		
	M0	M60	M90
30	2750	1789	531
70	2688	1682	431
90	2341	765	283

Figure 20. SPD rate of the composites stressed by a homopolar impulse superimposed DC voltage in the first 15s

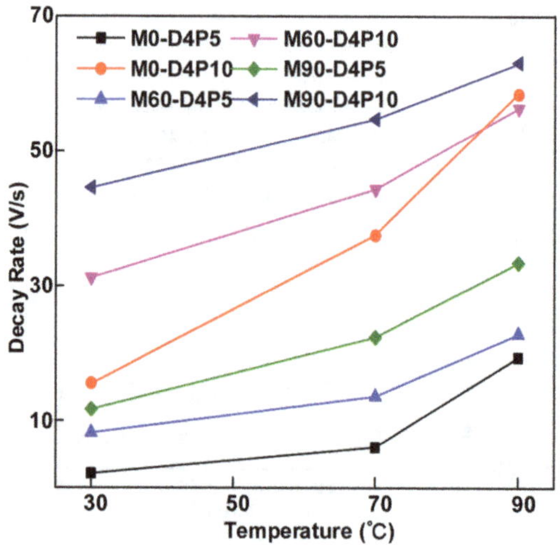

better in the sample after mixing with the sic particles. This phenomenon can be explained by the diffusion effect associated with nonlinear conductivity. When the rise of pulse voltage brings extra charge, the nonlinear conductivity of M60 and M90 starts to work, and a large number of newly arrived charges will spread to the adjacent area in a short time. As a result, the ISP increase of M90 sample was the smallest. In addition, the temperature can similarly suppress the surface charge, which is because all samples are within the range of positive conductance temperature coefficients (Du, 2019).

Surface Potential Property of the Composites Under Heteropolar Impulse Superimposed DC Voltage

Figure 21 shows the stress on ISP of the composites caused by the DC voltage superimposed by heterogeneous pulse. Similar to the previous section, the temperature can be reduced by the stress voltage to reduce the ISP due to its non-linear conductivity. SiC plays the same role for ISP. Samples with high levels of SiC have a more pronounced effect on surface charge, and more charge can be easily transferred to the surface. However, the different pole voltage does not seem to cancel the negative charge caused by the DC voltage. The ISP of all samples has an increasing trend, which is affected by the polarity voltage. Furthermore, for M0, the increase in ISP at 90°C is slightly greater than the increase at 30°C and 70°C, while the increase in ISP at 60°C is slightly greater than at low temperatures. Figure 22 shows the corresponding SPD rate for the first 15 s of the heteropolar pulse voltage. Temperature and SiC content are still effective in promoting surface charge decay of highly conductive materials. At the same time, the non-uniform pulse voltage exacerbates the accumulation of surface charge while slowing down the decay process. The results show that all heterogeneous pulse voltage stress samples have high ISP and low surface charge decay rate. Obviously, the SPD characteristic is caused by the non-linear conductivity. This phenomenon can be explained by the hypothetical model in Figure 23. When the heterogeneous pulse voltage reaches the end of the charging process, the polarity of the voltage changes suddenly, and many positive charges are injected into the sample surface. The hybrid charge is neutralized immediately. But in a short time this process cannot be completed. The positive charge is trapped in the area below the needle electrode and forms a positive charge area. As a result, the attraction between hybridizations has been around until the neutralization is complete. The surface charge in the adjacent region is attracted to the central position, which results in the increase of ISP detected. The attractive motion towards the center is also affected by conductivity. The charge is more likely to accumulate in the central region due to the high conductivity. Ncutralization then reduces the accumulation of positive charges and weakens the electric field. The surface potential starts to decay, but as long as the neutralization is not completed, the surface potential will still be greatly limited by the attraction. The decrease in the attenuation rate shown in Figure 22 corresponds to this. This explains the phenomenon of figure 21 at 90°C. Neutralization occurs during SPD. When the nonlinear conductivity is not large enough, the gravitational force can even offset the rapid decay rate caused by the nonlinear conductivity, and the attenuation rate is attenuated to the level of the sample not doped with SiC. The decrease of attenuation rate of M60 and M90 at 30°C shows the same trend as the above analysis. M0 without nonlinear conductance decays slowly under DC voltage. Therefore, there is no similar decrease in SPD rate.

The model in Figure 23 can be further inferred. The surface potential can be lowered when the pulse voltage is sufficiently strong because the negative charge accumulated under the direct current voltage can be immediately neutralized by a large amount of positive charges. In this case, the time required to move the sample from the needle electrode to the probe is shorter than the possible duration of strong attraction. The attraction effect can be detected

Figure 21. ISP of the composites stressed by a heteropolar impulse superimposed DC voltage

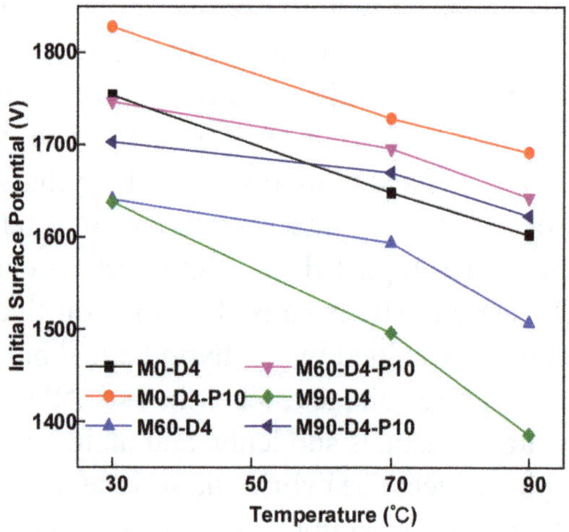

Figure 22. SPD rate of the composites stressed by a heteropolar impulse superimposed DC voltage in the first 15 s

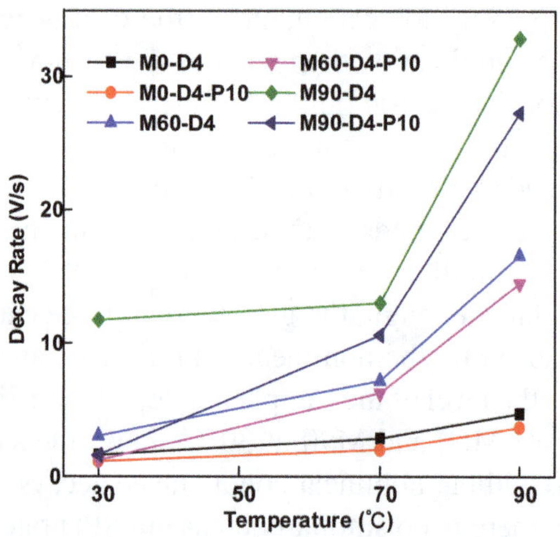

Figure 23. Surface charge movement stressed by a heteropolar impulse superimposed DC voltage

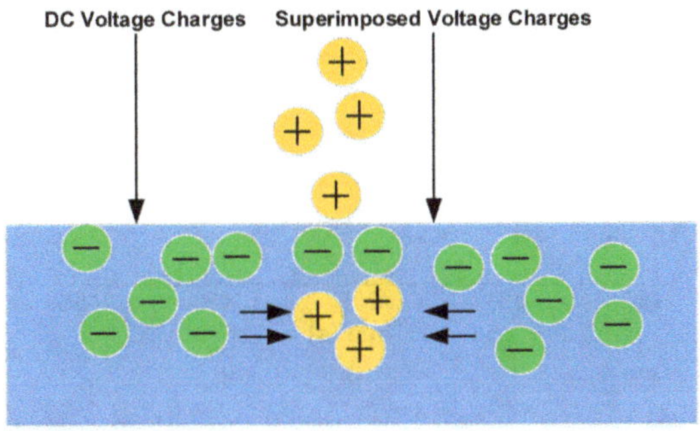

by an increase in the surface potential of the central region. Figures 24 and 25 show additional experimental data based on the above reasoning. The starting point of the horizontal axis is the center point below the electrode, and the number on the axis corresponds to the distance between the detecting point and the center point. Obviously, the surface potential of the sample increases in the central region, and the decrease in the surrounding region clearly shows this attraction force under the action of D4-P10. The surface potential of all regions will decrease if further stress d4-p15 is applied to

Figure 24. ISP of M60 at 70°C stressed by different voltages

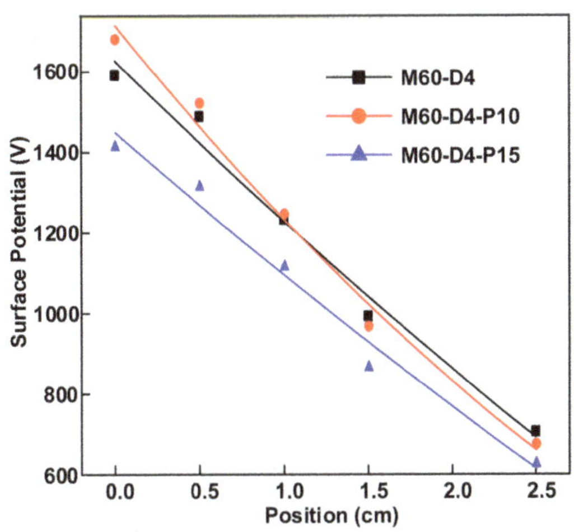

Figure 25. SPD rate of M60 at 70°C stressed by D4-P15

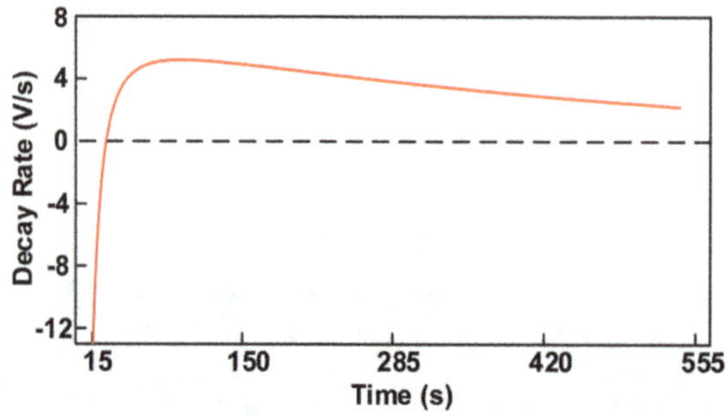

the sample. Figure 25 shows the attenuation rate of D4-P15 to M60 at 70°C. The surface potential increases in the first 15 seconds and then decays. These data fit well with the model in Figure 23.

CONCLUSIONS

In the whole temperature range, the nonlinear conductivity of SiR/SiC composite is obtained. Field strength and SiC content have a positive effect on the increase in conductivity. When the temperature increases, the threshold field decreases. At high SiC content, this phenomenon is more obvious. Carrier mobility and conductivity are more affected by temperature than the applied voltage. The nonlinear conductance increases as the surface potential generated by the repetitive pulse increases, causing the surface charge to dissipate immediately. Besides, this is more pronounced in samples with high SiC content. The effect of the increase in DC voltage on the accumulation of surface charge is more pronounced. The charge can be moved to the surface of the sample under continuous driving of a long-term DC voltage, while a repetitive pulse without a peak of the pulse cannot provide sufficient energy for the charge. With the increase of homopolar pulse voltage, ISP of all samples will increase. Obviously, the increase of SiC content and pulse voltage can accelerate the attenuation process. Non-uniform pulse voltages can exacerbate the accumulation of surface charge and slow down the SPD. The hybrid charge is neutralized immediately. However, it is difficult to complete this process in a short time. The positive charge is trapped in the region, and the attraction between the hetero-charges can explain the observed phenomenon.

REFERENCES

Ando, Y., & Itoh, T. (1987). Calculation of Transmission Tunneling Current Across Arbitrary Potential Barriers. *Journal of Applied Physics*, *61*(4), 1497–1502. doi:10.1063/1.338082

Boggs, S., Damon, D. H., Hjerrild, J., Holboll, J. T., & Henriksen, M. (2015). Effect of insulation properties on the field grading of solid dielectric dc cable. *IEEE Transactions on Power Delivery*, *16*(4), 456–461. doi:10.1109/61.956720

Castellon, J., Nguyen, H. N., Agnel, S., Toureille, A., Frechette, M., Savoie, S., ... Schmidt, L. E. (2011). Electrical Properties Analysis of Micro and Nano Composite Epoxy Resin Materials. *IEEE Transactions on Dielectrics and Electrical Insulation*, *18*(3), 651–658. doi:10.1109/TDEI.2011.5931049

Das-Gupta, D. K. (1990). Decay of Electrical Charges on Organic Synthetic Polymer Surface. *IEEE Transactions on Dielectrics and Electrical Insulation*, *25*(3), 503–508. doi:10.1109/14.55723

Donzel, L., Greuter, F., & Christen, T. (2011). Nonlinear Resistive Electric Field Grading Part 2: Materials and Applications. *IEEE Electrical Insulation Magazine*, *27*(2), 18–29. doi:10.1109/MEI.2011.5739419

Du, B. X., Yang, Z. R., Li, Z. L., & Li, J. (2018). Temperature-dependent nonlinear conductivity and carrier mobility of silicone rubber/SiC composites. *IEEE Transactions on Dielectrics and Electrical Insulation*, *25*(3), 1080–1087. doi:10.1109/TDEI.2018.006942

Du, B. X., Yang, Z. R., Li, Z. L., & Li, J. (2019). Nonlinear conductivity and charge transport characteristics of silicone rubber/SiC composites under impulse superimposed on DC voltage. *IEEE Transactions on Dielectrics and Electrical Insulation*, *26*(3), 776–783. doi:10.1109/TDEI.2019.8726024

Du, B. X., Yang, Z. R., Li, Z. L., & Li, J. (2019). Temperature-dependent charge property of silicone rubber/SiC composites under lightning impulse superimposed DC voltage. *IEEE Transactions on Dielectrics and Electrical Insulation*, *26*(3), 810–817. doi:10.1109/TDEI.2019.8726028

Du, B. X., Zhu, W. B., & Li, X. L. (2017). Effects of Direct Fluorination on Charge Coupling Behavior of Oil-Paper Insulation under DC and Pulse Voltages. *IEEE Transactions on Dielectrics and Electrical Insulation*, *24*(2), 947–955. doi:10.1109/TDEI.2017.006287

Gildenblat, G. S., Rao, A. R., & Cohen, S. S. (1987). Space-Charge-Limited Currents in Materials with Nonlinear Velocity-Field Relationships. *IEEE Transactions on Electron Devices*, *34*(10), 2156–2172. doi:10.1109/T-ED.1987.23212

Grundmann, M. (2010). *The Physics of Semiconductors: An Introduction Including Nanophysics and Applications* (2nd ed.). New York: Springer. doi:10.1007/978-3-642-13884-3

Hedvig, P. (1977). *Dielectric Spectroscopy of Polymers*. Budapest: Akademiai Kiadö.

Hjerrild, J., Boggs, S., Holboll, J. T., & Henriksen, M. (2001). DC-Field in Solid Dielectric Cables under Transient Thermal Conditions. *Proceedings of the 20001 IEEE 7th International Conference on Solid Dielectrics*: 58-61. 10.1109/ICSD.2001.955511

Hu, H., Wang, T., Zhang, H., & Song, J. (2010). Analytical Model for Intrinsic Carrier Concentration in (101)-Biaxially Strained Si. *International Conference on Information Science and Engineering*, 1864–1866. 10.1109/ICISE.2010.5691930

Jacobson, B., Jiang-Hafner, Y., Rey, P., Jeroense, G. M., Gustafsson, A., & Bergkvist, M. (2006). HVDC with Voltage Source Converters and Extruded Cables for up to ±300 kV and 1000 MW. *IEEE International Conference on Large High Voltage Electric Systems (ICLHVES).*

Lan, L., Zhong, Q., Yin, Y., & Li, X. (2013). Effect of Surface Fluorination on Space Charge Behavior at LDPE/EPDM Interface. *2013 IEEE International Conference on Solid Dielectrics (ICSD)*, 444-447. 10.1109/ICSD.2013.6619657

Li, J., Du, B. X., & Hang, H. (2017). Suppressing Interface Charge Between LDPE and EPDM for HVDC Cable Accessory Insulation. *IEEE Transactions on Dielectrics and Electrical Insulation*, *24*(3), 1331–1339. doi:10.1109/TDEI.2017.006073

Llovera, P., & Molinie, P. (2004). New Methodology for Surface Potential Decay Measurements: Application to Study Charge Injection Dynamic on Polyethylene Films. *IEEE Transactions on Dielectrics and Electrical Insulation*, *11*(6), 1049–1056. doi:10.1109/TDEI.2004.1387828

Nelson, P. N., & Hervig, H. C. (1984). High Dielectric Constant Materials For Primary Voltage Cable Terminations. *IEEE Transactions on Power Apparatus and Systems*, *11*(103), 3211–3216. doi:10.1109/TPAS.1984.318558

Perlman, M. M., Sonnonstine, T. J., & St. Pierre, J. A. (1976). Drift Mobility Determinations Using Surface Potential Decay in Insulators. *Journal of Applied Physics*, *47*, 5016–5021. doi:10.1063/1.322511

Simmons, J. G., & Tam, M. C. (1973). Theory of Isothermal Currents and the Direct Determination of Trap Parameters in Semiconductors and Insulators Containing Arbitrary Trap Distributions. *Physical Review B: Condensed Matter and Materials Physics*, *7*(8), 3706–3713. doi:10.1103/PhysRevB.7.3706

Strumpler, R., & Glatz-Reichenbach, J. (1999). Conducting Polymer Composites. *Journal of Electroceramics*, *3*(4), 329–346. doi:10.1023/A:1009909812823

Tanaka, T. (2005). Dielectric Nanocomposites with Insulating Properties. *IEEE Transactions on Dielectrics and Electrical Insulation*, *12*(5), 914–928. doi:10.1109/TDEI.2005.1522186

Tanaka, T., Kozako, M., Fuse, N., & Ohki, Y. (2005). Proposal of a Multi-Core Model for Polymer Nanocomposite Dielectrics. *IEEE Transactions on Dielectrics and Electrical Insulation*, *12*(4), 669–681. doi:10.1109/TDEI.2005.1511092

Virsberg, L., & Ware, P. H. (1967). A New Termination for Underground Distribution. *IEEE Transactions on Power Apparatus and Systems*, *86*(9), 673–681.

Vu, T. T. N., Teyssedre, G., Vissouvanadin, B., Roy, S. L., & Laurent, C. (2015). Correlating Conductivity and Space Charge Measurements in Multi-Dielectrics under Various Electrical and Thermal Stresses. *IEEE Transactions on Dielectrics and Electrical Insulation*, *22*(1), 117–127. doi:10.1109/TDEI.2014.004507

Wang, Z., Nelson, J. K., Hillborg, H., Zhao, S., & Schadler, L. S. (2012). Nonlinear conductivity and dielectric response of graphene oxide filled silicone rubber nanocomposites. *2012 Annual Report Conference on Electrical Insulation and Dielectric Phenomena (CEIDP)*, 40-43. 10.1109/CEIDP.2012.6378717

Xie, J. C., Hu, J., He, J. L., Guo, Z. C., & Yin, Y. (2013). Effect of Silicone Rubber Polymer Composites on Nonuniform Electric Fields of Rod-plane Gaps. *IEEE Conference on Electrical Insulation and Dielectric Phenomena*, 399-402. 10.1109/CEIDP.2013.6747455

Yang, X., Hu, J., Chen, S., & He, J. L. (2016). Understanding the percolation characteristics of nonlinear composite dielectrics. *Scientific Reports*, *6*(1), 30597. doi:10.1038rep30597 PMID:27476998

Zhou, Y., He, J. L., Hu, J., Huang, X. Y., & Jiang, P. K. (2015). Evaluation of Polypropylene/Polyolefin Elastomer Blends for Potential Recyclable HVDC Cable Insulation Applications. *IEEE Transactions on Dielectrics and Electrical Insulation*, *22*(2), 673–681. doi:10.1109/TDEI.2015.7076762

About the Author

Boxue Du received the M.E. degree in electrical engineering from Ibaraki University and the Ph.D. degree from Tokyo University of A&T. Since 2002 he has been a Professor and Director-founder of the Institute of High Voltage at the School of Electrical and Information Engineering, Tianjin University, China. His research interests are focused on dielectric failure mechanisms of polymer insulating materials, electrical insulation technology and application of polymer dielectrics under various extreme environments such as cryogenic, high temperature, high altitude, gamma-ray irradiation and high-intensity magnetic field. He has published 4 books and 8 book chapters in Polymer Dielectrics, and authored about 500 papers and over 150 of them published in IEEE Transactions. He has also served industry and start-ups as a consultant, i.e. an IEEE China Council Council on Superconductivity Chapter, Chief Committee Member; Chinese Society of Electrical Engineering, High Voltage New Technology Committee Member. He is a Fellow of IET, senior member of IEEE and CSEE, Committee Member at several WG in CIGRE and IEEE Standards. He is an Editorial Boards Member of IET High Voltage; Chinese Journal of High Voltage Engineering. He is an Associate Editor of the IEEE Transactions on Dielectrics and Electrical Insulation, an Associate Editor of IEEE Access and a Subject Editor of the IET Nanodielectrics.

Index

G

H

I

L

M

N

O

P

S

T

U

V

W

www.ingramcontent.com/pod-product-compliance
Lightning Source LLC
LaVergne TN
LVHW081520300926
849161LV00006B/1233